Holographic
Microscopy
of
Phase Microscopic Objects
Theory and Practice

Holographic Microscopy

of

Phase Microscopic Objects

Theory and Practice

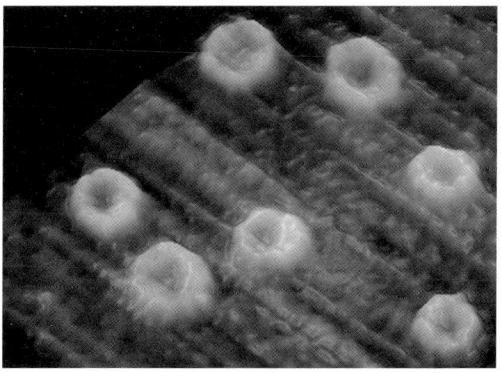

Tatyana Tishko
Tishko Dmitry
Titar Vladimir

V N Karazin Kharkiv National University, Ukraine

World Scientific

NEW JERSEY · LONDON · SINGAPORE · BEIJING · SHANGHAI · HONG KONG · TAIPEI · CHENNAI

Published by

World Scientific Publishing Co. Pte. Ltd.

5 Toh Tuck Link, Singapore 596224

USA office: 27 Warren Street, Suite 401-402, Hackensack, NJ 07601

UK office: 57 Shelton Street, Covent Garden, London WC2H 9HE

British Library Cataloguing-in-Publication Data
A catalogue record for this book is available from the British Library.

HOLOGRAPHIC MICROSCOPY OF PHASE MICROSCOPIC OBJECTS
Theory and Practice

Desk Editor: Tjan Kwang Wei

ISBN-13 978-981-4289-54-2
ISBN-10 981-4289-54-X

Printed in Singapore by Mainland Press Pte Ltd.

This book is devoted to Teachers of ours

Preface

Holography was born as a "new microscopy principle". Holographic microscopy is an important practical application of holography. It has a lot of unique advantages which open new possibility in study of phase microscopic objects.

Application of holographic and digital methods in light microscopy has made it possible not only to obtain 3D images of untreated cells but also to measure their morphological parameters. The proposed book contains either theory, or practice of classical and holographic microscopy of phase microscopic objects.

The main attention is paid to practical applications. Separate chapters of the book are devoted to study of 3D morphology of human blood erythrocytes and thin films.

We think that the book will be of some interest for physicists, physicians and students of the corresponding specialties.

Tatyana Tishko

Contents

Introduction

A lot of objects of our world are microscopic objects. They are invisible for human eye. Since the time of Romans it was realized that glasses of certain shapes could magnify objects. Lenses were being used as corrective eyeglasses. But the history of microscopy started on late 1500s. The first simple microscopes were developed by Dutch spectacle maker Antony von Leeuwenhoek (1632-1723). A. Leeuwenhoek built microscopes with a single small lens which reached magnification up to 270x. Using single lens microscopes, A. Leeuwenhoek was able to describe organisms and tissues, such as bacteria and red blood cells, which were not known to exist. The Englishman Robert Hooke (1635-1703) is known to have used microscopes consisting of an objective lens and an eyepiece for the first time. Since that time microscopes are used for investigation of microscopic objects. But by the middle of nineteenth century there was not a theory if microscopic imaging though significant improvement had been made in the light microscope design. Ernst Abbe (1840-1905) was the first to apply physical principles to the problem of image formation by microscope. His theory of microscopic imaging is based on the idea that the microscopic image is the result of interference of the waves diffracted on the microscopic object and the zero-order wave [1, 2].The theory has explained the dependence of microscope resolution from numerical aperture of microscope objective. E. Abbe's theory has stimulated development of difference microscopy methods and devices.

The particular group of microscopic objects is presented by phase microscopic objects. Phase microscopic objects do not change the intensity of the radiation transmitted through them, they insert only phase increments and are inaccessible to direct observation by optical microscope. They are cells and tissues of living organism, crystals, thin films, etc. To observe such phase microscopic objects it is necessary to convert phase changes inserted by them into the light wave passed through them into intensity changes. Though, special methods must be used for phase microscopic objects study. The two main methods were proposed for phase microscopic objects visualization: phase-contrast and interference contrast methods. For the first time the problem of realizing phase-contrast was solved by F. Zernike in 1934 [41, 4]. For the phase-contrast method and the phase-contrast microscope he won Nobel Prize in Physics in 1953. The first interference- contrast microscope was described by M. Sagnas in 1911 [4]. Since there a lot of interference microscopes were proposed. They mainly differ by the method of beam splitting and combining.

   Phase-contrast microscopes are widely used for phase microscopic objects visualization. But they allow only two-dimensional visualization of phase microscopic objects, measurement of their geometrical parameters is impossible. Interference microscopes allow quantitative measurements, but they are mainly used only for thin films thickness measurements. The problem of 3D visualization of phase microscopic objects has not been set and solved in classical microscopy. Till recently, electron microscopy was the only method for the 3D imaging of microscopic objects. However, this high-resolution method is "destructive". Long preliminary treatment of the sample is needed; this renders study of native cells impossible. In addition, such a treatment affects cells, and the reliability of the results of electron microscopy can be debated.

   Improving resolution of electron microscope D. Gabor proposed the method of holography in 1948 (Nobel Prize in Physics in 1971) [5]. Though Gabor's idea has not realized in microscopy, the advent and development of holography as the method for recording and reconstruction of the wave phase and wave amplitude resulted in the appearance of holographic analogs of the classical microscopy methods for phase microscopic object study. They are holographic phase-contrast method [20, 22], the method of holographic microinterferometry [21, 6], and polarization-contrast method [19, 20, 23]. The first holographic interference microscope, realizing the methods was created at the

Laboratory of Holography, Kharkov National University, Ukraine. For the first time holographic addition and subtraction of waves in an interference fringe for obtaining images of phase microscopic objects was realized. The contrast images of blood erythrocytes in a bright and dark interference fringes and their interferograms were obtained using the microscope.

Application of the holographic methods has opened new possibility in microscopy of phase microscopic objects. Now it was made possible not only to visualize phase microscopic objects, but also to measure their geometrical parameters using their interferograms.

The development of computers and the methods for the digital data processing has led to a new stage in light microscopy. The problem of 3D visualization of phase microscopic objects has been solved by combining the holographic methods with the methods for digital image processing. The first digital holographic interference microscope (DHIM) which allows the real-time 3D imaging of phase microscopic objects and the carrying out quantitative measurements of their geometrical parameters has been created on the base of the holographic interference microscope [24, 27, 32]. For the first time it has become possible to obtain 3D images of native cells. The first 3D images of the native human blood erythrocytes were obtained using the DHIM in 1998 [24].Lately a few digital holographic microscopes were proposed, in which digital reconstruction and processing a hologram of a microscopic objects is used for their 3D imaging [10, 14].

In this book we present the comparative analysis of the classical and holographic microscopy methods for phase microscopic objects visualization; description of the digital holographic interference microscope (DHIM), and the result of DHIM application for experimental study of the 3D morphology of biological and technical phase microscopic objects.

Chapter 1

Classical Microscopy and Methods of Phase Microscopic Objects Visualization

1.1 The theory of object imaging by microscope

The first satisfactory wave theory of image formation by microscope was formulated by E. Abbe in1873 [1, 2]. It is based on the idea that the microscopic image is the result of interference of the wave diffracted on the microscopic object and the zero-order wave (undeviated, undiffracted wave). Theretofore it was considered that image in microscopes are formed by the usual refraction laws, i.e. in the same way as they are formed in telescopes, cameras and etc. The dependence of microscope resolution from numerical aperture of microscope objective was also incorrectly explained.

Illuminated objects instead of self-luminous are viewed through a microscope. This means that different points of the object scatter the waves falling on them from the same point of the light source; and, consequently, the light propagating from different point of the object occurs to be coherent. According to Abbe, the object is similar to a diffraction grating. So that not only every element of aperture of the objective, but also every element of the object must be taken into account in determining the complex disturbance at any particular point in the image plane.

Figure1 shows the schematic diagram to illustrate Abbe's theory of image formation.

Let a grating object situated in the object plane OO is illuminated coherently by a plane light wave and observed using microscope objective L. In this case the period of the grating d is the characteristic

5

detail, and the resolution of the microscope determines the minimal value of d .Diffraction of the plane wave on such object results in forming a diffraction pattern in the back focal plane FF of the microscope objective. A number of diffraction maximums are appeared (0, 1, 2);

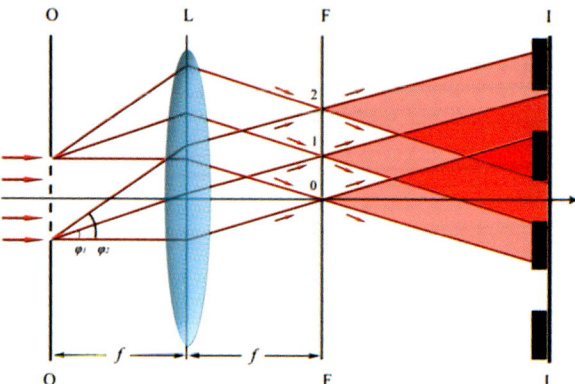

Fig.1. Schematic diagram to illustrate Abbe's theory of image formation. OO- object plane; L- objective; FF-back focal plane of the objective; II- image plane; f-focal distance of the objective.

Angular distances (φ_1, φ_2) between them are determined by the period of the grating. Positions of these maximums are given by the equation

$$d \sin \varphi = m\lambda , \tag{1}$$

where m is the integral number which determines the orders of the maximums, λ is the wavelength of the light being used. As it follows from Eq. (1), the zero maximum ($m=0$) is on the optical axis, the first-order maximums are situated in the directions which are determined by the relations

$$\sin \varphi_1 = \pm \frac{\lambda}{d} ; \tag{2}$$

the second-order maximums in the directions

$$\sin \varphi_2 = \frac{2\lambda}{d} ; \tag{3}$$

etc. Every point in the back focal plane can be considered as a source of secondary disturbance. All these diffraction maximums correspond to coherent rays, so, the light waves from all these secondary sources interfere with each other and form image of the object in the image plane II of the objective.

The non-diffracted zero-order rays and the first- and higher-order rays are spatially separated in the back focal plane *FF* of the microscope objective, but they are combined in the image plane *II* .

So, either the diffraction maximums, or the final picture in the image plane depend on the object and can be considered as its image. The pattern in the focal plane of the objective *FF* is called primary image (the spectrum of the object) and the pattern in the image plane *II* is called secondary image (the image of the object).To obtain the exact image of the object all diffraction spectrum have to take part in image formation. But this is impossible because of finite size of the microscope objective. At least two diffraction orders (the zero and the first ones) have to be passed through the optical system in order that the interference occurs and the object image is formed. If the spatial frequencies of the object are so high and aperture is so small that no orders of diffraction spectrum from fine details of the object are transmitted by the aperture of the microscope objective, then that detail will be invisible no matter what magnification is used.

The possibility of the microscope objective to collect the diffracted light is characterized by the numerical aperture *NA*. The numerical aperture is defined by

$$NA = n\sin\frac{u}{2},\tag{4}$$

where n is the index of refraction of the medium in which the objective is working, and $\frac{u}{2}$ is the half angle of the maximum corner of light that can enter the objective. This quantity indicates the resolving power of the microscope. Though, the minimal resolved element of an object is determined by the formula:

$$d \geq \frac{\lambda}{NA},\tag{5}$$

Using immersion liquid, it is possible to increase the quantity of the numerical aperture to 1.35; and the minimal resolvable distance between two elements of the microscopic object under study can be about half of the wavelength of the radiation being used.

1.2 Phase microscopic objects

When a transparent microscopic object, which is situated in the plane (x, y), is illuminated by the plane monochromatic wave, then the transmittance function $F(x, y)$ of the object is defined by the relation

$$F(x, y) = \frac{V(x, y)}{V_0(x, y)}, \qquad (6)$$

where $V_0(x, y)$ is the wave in the absence of the microscopic object; $V(x, y)$ is the wave in the present of the microscopic object. In general this function is complex, because the object can insert amplitude and phase changes into the light wave transmitted through it. When the microscopic object changes only the amplitude of the wave, i.e. arg $F = 0$, one can speak about the "amplitude" microscopic object. If the microscopic object change only the phase of the wave transmitted through it, i.e. $|F| = 1$, then we speak about "phase" microscopic object. So, the transmittance function of the phase microscopic object has the form:

$$F(x, y) = \exp[i\varphi(x, y)] , \qquad (7)$$

where $\varphi(x, y)$ is the phase increment inserted by the microscopic object into the wave transmitted by it. If the phase microscopic object is observed through a microscope objective, then the disturbance in an arbitrary point of the image plane [2]:

$$V(x', y') = CF(x, y), \qquad (8)$$

where $F(x, y)$ is the transmittance function of the microscopic object. Then

$$V(x', y') = C \exp[i\varphi(x, y)]. \qquad (9)$$

The intensity in the image plane is equal

$$I(x', y') = |V(x', y')|^2 = C^2, \qquad (10)$$

i.e. the intensity in the image plane is not modulated by the phase relief of the object and the phase object is invisible when it is observed through a microscope.

Live cells are colorless and transparent microscopic objects. To visualize such micro-objects specimen staining is commonly used in biology and medicine but this involves killing and fixing the sample.

Staining may also introduce artifacts, apparent structural details that are caused by the processing of the specimen.

The main drawback of non-optical methods of studying phase microscopic objects is that an effect is produced on the microscopic object itself.

Another approach to the study of phase microscopic objects consists of producing an effect not on the test specimen but on a light wave transmitted through it.

To visualize phase microscopic objects, it is necessary to convert the phase increments inserted by them into the transmitted light wave into intensity changes. Interference is the phenomenon in which the phase difference of two interacting waves can be converted into changes of the resulting intensity. Two main methods based on the interference phenomenon are used in classical microscopy for the phase microscopic objects visualization. They are F. Zernike phase-contrast method and interference contrast method.

1.3 Interference phenomenon

In physics, interference is the superposition of two or more waves that result in a new wave pattern. Interference usually refers to the interaction of waves which are correlated or coherent with each other, either because they come from the same source. Two waves are said to be coherent if they have a constant relative phase. Interference is the phenomenon in which phase differences of two waves are converted into intensity changes. The interference can be of two types: in an infinitely wide fringe and in fringes of finite width. The first type of interference takes place, when the interacting waves are propagating in the same direction. In this case we obtain the resulting pattern with the homogeneous intensity distribution. The main equation of interference of two coherent linearly polarized waves $A_1(x, y) = a_1(x, y)\exp[-i\phi_1(x, y)]$ and $A_2(x, y) = a_2(x, y)\exp[-i\phi_2(x, y)]$ has the form [2]:

$$I = a^2_{\,1} + a^2_{\,2} + 2a_1 a_2 \cos(\phi_2 - \phi_1)\cos\beta, \qquad (11)$$

or

$$I = I_1 + I_2 + 2\sqrt{I_1 I_2}\cos(\phi_2 - \phi_1)\cos\beta \qquad (12)$$

where a_1, a_2 are amplitudes of the waves; I_1, I_2 are the intensities of the waves, ϕ_1, ϕ_2 are the phase of the waves, β is the angle between the direction of linear polarization of the waves. $I_1 = a^2_1$; $I_2 = a^2_2$. If the interacting waves have orthogonal linear polarizations, i.e. $\beta = 90^0$, then the interference of the waves does not occur. If the waves have equal intensity, and polarizations i.e. $I_1 = I_2 = I_0$, $\beta = 0$, then

$$I = 2I_0[1 + \cos(\phi_2 - \phi_1)] . \qquad (13)$$

So, the resulting intensity is modulated by the phase difference of the waves. Change of the phase difference of the wave results in changing the resulting intensity. Though, we can obtain increase or decrease of the resulting intensity. When $\phi_2 - \phi_1 = 2\pi m$, where $m = 0, 1, 2...$, then the resulting intensity is maximal and is equal to $I_{max} = 4I_0$. If $\phi_2 - \phi_1 = (2m + 1)\pi$, the resulting intensity is equal to $I_{min} = 0$; so the waves quench each other. The second type of interference occurs when there is some angle θ between the waves. In this case instead of homogeneous field, a line field is obtained, with uniform alternation of intensity maxima and minima. The interference equation has the form:

$$I(x, y) = 2I_0\{1 + \cos[(\phi_2 - \phi_1) - \frac{2\pi x}{T}]\} , \qquad (14)$$

where T is the period of the system of interference fringes (the distance between two intensity maximum or minimum in the interference pattern); (x, y) is the plane of observation. The period of the interference pattern T is equal:

$$T = \frac{\lambda}{2\sin(\theta/2)} , \qquad (15)$$

where λ is the wavelength of the light being used. The period of the interference pattern is decreasing under increasing the angle between the waves.

1.4 F. Zernike's phase-contrast method

Using E. Abbe's theory as a basis, F. Zernike proposed a method of phase microscopic object visualization that became known as phase-contrast method [41, 4].

The essence of the method is that it is possible independently affect the diffracted and direct waves changing their phase relationship. As it follows from the theory of image formation by a microscope, the diffracted wave is shifted in phase by $\frac{\pi}{2}$ relative to the phase of the zeroth-order wave and has low intensity by comparison with the intensity of the zeroth-order wave. So, a special phase plate that retards or advances the phase of the zeroth-order wave by $\frac{\pi}{2}$ is placed in the microscope in the zeroth order of diffraction. As a result, a phase difference equal 0 or π is created between the zeroth order wave and the diffracted wave. The waves in the former case are in phase (addition of waves), and this strengthens the intensity of the image by comparison with the background, whereas they are anti-phased in the latter case (subtraction of waves), and this weaken the intensity of the image by comparison with the background.

The effect of the phase plate, situated in the back focal plane of the microscope objective, is described by the transmission function:

$$A = ae^{i\alpha}, a \le 1 \qquad (16)$$

where a is the transmittance coefficient of the phase plate, α is the phase increment inserted by this plate. In this case the intensity in the image plane of microscope has the form [2]:

$$I(x', y') = |c|^2 \left\{ a^2 + 2\left[1 - a\cos\alpha - \cos\varphi(x, y) + \cos(\alpha - \varphi(x, y))\right] \right\} \quad (17)$$

where c is a constant. If $\varphi(x, y)$ is small, the phase increment inserted by the plate $\alpha = \pm\frac{\pi}{2}$, then the intensity $I(x', y')$ in the microscopic object image in the image plane of the microscope is modulated by the phase increment $\varphi(x, y)$ inserted by the microscopic object into the transmitted wave:

$$I(x', y') = |c|^2 [a^2 \pm 2a\varphi(x, y)] \qquad (18)$$

where $|c|^2$ is the intensity of the zero order wave.

So, the phase increments, inserted by the microscopic object into the light wave transmitted through it, are converted into intensity changes in its image, which are proportional (with an accuracy up to the constant) to the phase increments. If the phase of the zeroth-order wave lags behind the phase of the diffracted wave (the sign "+" in Eq. (18)), then the image of the microscopic object is brighter than the background. This is so called *bright phase contrast*. It the phase of the zeroth- order wave leads phases of diffracted waves (the sign "-" in Eq. (18)), then the image of the microscopic object is darker than the background. This is so called *dark phase contrast*.

The contrast of the image is determined by the ratio of the second term in the brackets of Eq. (18) to the first, and it is equal $\dfrac{2\varphi(x, y)}{a}$. It follows that the contrast of the image is higher, the higher the absorption of the plate. So, the satisfactory contrast is achieved by absorption of the direct light, and as a result, the decrease of total intensity takes place.

Thus, in Zernike's phase contrast method, the phase increments inserted by a microscopiv object into a wave transmitted through it are converted by means of interference into an intensity change of its image. However, since waves of different intensity interact in the phase-contrast method, adding and subtracting the waves does not provide high image contrast, and consequently the method does not achieve high sensitivity. Absorbing phase plates are used to increase the image contrast, increasing the sensitivity of the method while reducing the overall illumination.

The phase-contrast method is based on the possibility of separate acting on the direct light without changing the light diffracted at the microscopic objcct. Consequently, it can only be used for observing small objects, in the other case the direct light will be mixed with the diffracted light.

Thus, the phase-contrast method operates the more successfully, the smaller the phase microscopic object in optical thickness and dimensions.

1.5 The interference-contrast method

This method is also based on the phenomenon of interference of two waves. The interference- contrast method can be realized in two main variants: interferometry in an infinitely wide fringe and interferometry in fringes of finite width. In the first case the phase microscopic object becomes visible due to intensity changes, and in the second case due to displacements of interference fringes. In the interference –contrast method case phase microscopic objects become visible due to the interference of two waves of equal intensity, one of which passes through the microscopic object under study. In two-beam interferometer two compared waves are formed simultaneously, but propagate in different ways, then are combined and interfere. Figure 2 shows the principle layout of the two-beam interferometer.

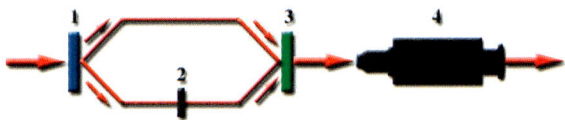

Fig.2. The principle layout of the two-beam interferometer. *1*- beam splitter; *2*- microscopic object; *3*- beam splitter; *4*- microscope.

The beam of light, outgoing from the source of light, is split by the beam splitter *1* in two beams. One beam passes through the microscopic object *2*, whereas the second beam passes by it. The beams are combined by the beam combiner *3* and form the single beam which is directed to the microscope *4*. There are a few types of classical interference microscopes (microinterferometers). They generally differ by the methods of separation and combining the light beams; every interferometry method requires a special device.

1.5.1 *Interferometry in an infinitely wide fringe*

The interferometry in an infinitely wide fringe solves the same problem as the phase-contrast microscopy. In the method of the interferometry in an infinitely wide fringe the angle between the combined beams is equal to 0. Thus, the interfering waves propagate in the same direction. The resulting image intensity I_{im} in the case of interference of two coherent waves of equal intensity passed two identical channels of the interferometer and which differ only by the phase increment $\varphi(x, y)$ inserted by the microscopic object under study is determined by the main interference equation (11):

$$I_{im}(x, y) = 2I_0\{1 + \cos[\Delta\phi + \varphi(x, y)]\}.\qquad(19)$$

And background intensity I_b is equal to:

$$I_b(x', y') = 2I_0[1 + \cos\Delta\phi].\qquad(20)$$

Here I_0 is the intensity of each of the interacting waves; $\Delta\phi$ is the phase difference of the waves; $\varphi(x, y)$ is the phase increment inserted by the microscopic object under study into the wave transmitted through it.

It follows from Eqs. (19) and (20) that the image intensity differs from the background intensity and depends on the phase increment inserted by the microscopic object.

Let us estimate the effects of different phase increments on the intensities of the image and the background. Suppose that the interferometer is adjusted so that the phase difference of the interfering waves is equal 0, i.e., $\Delta\phi = 0$. That means that interacting waves are in phase (addition of waves).Then

$$\begin{aligned}I_{im+}(x', y') &= 2I_0\left[1 + \cos\varphi(x, y)\right],\\ I_{b+}(x', y') &= 4I_0.\end{aligned}\qquad(21)$$

So, as it seen from the equation, in the case of the wave addition one obtains dark images of microscopic objects against the bright background.

If the interferometer is adjusted so that $\Delta\phi = \pi$; it means that interfering waves are anti-phased (subtraction of waves).Then

$$\begin{aligned}I_{im-}(x', y') &= 2I_0\left[1 - \cos\varphi(x, y)\right],\\ I_{b-}(x', y') &= 0.\end{aligned}\qquad(22)$$

So, in the case of subtraction of the waves we obtain bright images of microscopic objects against the dark background.

Thus, the phase microscopic object is observed as dark on a bright background in the case of addition of the waves, whereas it is observed as bright on the dark background in the case of subtraction. The intensity in the image of the microscopic object is modulated by the phase increments inserted by the microscopic object into the transmitted wave. This method is equivalent to the phase-contrast method. However, although both methods are based on the interference of waves transmitted through a phase microscopic object and bypassing it, they have the difference that waves of equal intensity interact in the interference-contrast method. This makes it possible to maximize the image contrast and, consequently, the sensitivity of the method. Besides, the method is applicable for microscopic objects of every size.

However, the advantages of the interference-contrast method are reduced by the complexity of the apparatus and the difficulty of adjusting the interferometers. An essential problem in classical interferometry is the problem of obtaining two identical wave fronts. And this method is not used in classical microscopy for biological phase microscopic object imaging.

1.5.2 *The method of interferometry in fringes of finite width*

The method of interferometry in fringes of finite width involves the interaction of the waves between which there is a certain angle θ. In this case instead of homogeneous field, a line field is obtained, with uniform alternation of intensity maxima and minima. In this case in accordance with the main interference equation (11), the intensity distributions in the image of the microscopic object I_{im} and in the background I_b have the form:

$$I_{im}(x',y')=2I_0\left\{1+\cos\left[\Delta\phi+\varphi(x,y)-\frac{2\pi x}{T}\right]\right\},$$

$$I_b(x',y')=2I_0\left[1+\cos\left(\Delta\phi-\frac{2\pi x}{T}\right)\right].$$

(23)

where $T=\dfrac{\lambda}{2\sin(\theta/2)}$.

T is the period of the background interference pattern; I_0 is the intensity of each of the interacting waves, λ is the wavelength of the radiation being used; $\Delta\phi$ is the phase difference of the interacting waves, $\varphi(x, y)$ is the phase increment inserted by the microscopic object under study into the transmitted wave.

The equation of dark interference fringes has the form:

$$\Delta\phi + \varphi(x, y) - \frac{2\pi x}{T} = (2m+1)\pi .$$
(24)

And the equation of bright interference fringes is

$$\Delta\phi + \varphi(x, y) - \frac{2\pi x}{T} = 2m\pi .$$
(25)

where |m|=0,1,2,…

One can see that the phase increments, inserted by the microscopic object under study into the wave transmitted through it, result in change of the initial interference pattern, that is manifested in displacements of the interference fringes in the image of the object. These displaced interference fringes show a phase silhouette of the object.

As it follows from Eqs. (24) and (25), the displacement $h(x', y')$ of the image interference fringe in some point of the image plane is proportional to the phase increment $\varphi(x, y)$ inserted by the microscopic object, i.e.

$$h(x', y') = \frac{T\varphi(x, y)}{2\pi} .$$
(26)

So, measuring the displacements of the image interference fringes $h(x', y')$ and the period T of the background interference pattern, it is possible to determine the phase increments:

$$\varphi(x, y) = \frac{2\pi h(x', y')}{T} .$$
(27)

Thus, the method of interferometry in fringes of finite width gives the possibility not only to visualize the phase microscopic object, but also to measure the phase increments inserted by the microscopic object into the light wave transmitted through it, and to determine the thickness of the microscopic objects under study. But realization of the interferometry method in classical microscopy meets some difficulties. In the first place, this is the mentioned above difficulty of creation of two identical wave fronts, and, as a result, high requirements to the quality of optics, complexity of the adjustment of the microinterferometer. And practically the classical interference microscopes are used only for thin films thickness measurements.

1.6 The polarization-contrast method

All materials can be divided into two classes: isotropic and anisotropic materials. Isotropic materials, which include liquids, gases, glasses and cubic crystals, demonstrate the same optical properties in all directions. They have only one refractive index because the velocity of light is the same in all directions. The chemical bonds holding the material together are the same in all directions, so that the state of the polarization of the light passing through the material is not changed. Anisotropic materials which include the majority of all solid substances (for example, crystals, soft tissues etc.) have different optical and mechanical properties in different directions. Anisotropic materials change the polarization of light passing through them. This makes it possible to visualize anisotropic phase microscopic objects using the polarization- contrast method. In order to accomplish this task the microscope must be equipped by two polarizers. The first polarizer is positioned in the light path somewhere before the specimen. The second polarizer (which is called analyzer) is placed in the optical pathway between the microscope objective and observation camera. The contrast of the image arises from the interaction of the plane-polarized light entering from the polarizer with the optically anisotropic specimen. Figure 3 shows the layout of the polarization microscope.

Fig.3. The principle layout of the polarization microscope. *1* – polarizer; *2*- microscope; *3*- analyzer.

Optical anisotropy manifests itself basically in the form of birefringence (double refraction).Birefringent specimens act as beam splitters and divide a ray of plane-polarized light into two individual rays. One of the rays follows the ordinary refraction laws, and so it is called the ordinary ray. The second one does not obeys the law and is called the extraordinary ray. They are polarized in mutually perpendicular planes. The velocities of these rays are different and vary with propagation direction through the specimen. If the material has a single axis of anisotropy   (an axis of symmetry, such as normal to crystalline layers), the birefringence can be formalized by assigning two different refractive indices to the material for different polarizations. The birefringence magnitude is then defined by

$$\Delta n = n_e - n_0,\qquad(28)$$

where n_0 and n_e are, correspondently, the refractive indices for polarizations perpendicular (ordinary) and parallel (extraordinary) to the axis of anisotropy. So, it means that the two waves with mutually orthogonal polarizations propagating in one direction inside the crystal gain the phase difference due to the difference of the refractive indices. The phase difference $\Delta\varphi$ is equal

$$\Delta\varphi = \frac{2\pi}{\lambda}t\Delta n,\qquad(29)$$

where λ is the wavelength of the light being used, t is the thickness of the crystal. So, the phase difference depends on the sickness of the specimen.

These two waves can not interfere because they have orthogonal polarizations. But behind the linear analyzer, which transmits vibrations only in one plane, which is parallel to its transmittance plane, and makes these two waves plane-polarized in the same plane, interference of the ordinary and extraordinary waves occurs.

The intensity of the light transmitted through two polarizers is determined by the relation:

$$I = I_0 \cos^2 \alpha.\qquad(30)$$

where α is the angle between the transmittance axes of the polarizers. When the polarizer and the analyzer are crossed, i.e. $\alpha = \frac{\pi}{2}$ (their transmittance planes are orthogonal), no light is transmitted through the optical system in the absence of the specimen. When the anisotropic specimen is put in front of the microscope objective, its bright image, which is the result of interference of the ordinary and extraordinary waves, can be observed on the dark background. The intensity of the image depends on the phase difference $\Delta\varphi$. If the thickness of the specimen is nonuniform, then the intensity distribution in its image depends on the thickness distribution. So, the method of polarization contrast can be used for visualization of anisotropic microscopic objects. This method allows one to obtain images of anisotropic microscopic objects with maximal contrast. The method of polarization contrast is the method which improves the quality of images of anisotropic specimens when compared with other techniques such as phase-contrast, and interference- contrast methods, because the method makes it possible to remove all noisy isotropic microscopic objects which always present in the specimen (dust, scratches, defects of optics, etc.) and worsen the quality of the image of the microscopic object under study. The method allows obtaining the information about the structure of materials which can not be available with any other optical microscopy technique. Polarization microscopes are widely used in practice for anisotropic microscopic objects investigation.

The classical methods of phase microscopic objects visualization were considered in this chapter. They are phase-, interference- and polarization contrast methods. All these methods are based on the interference phenomenon. It was shown that phase-contrast method is optimal for small ant thin microscopic objects. The interference –contrast method has a few advantages in comparison with the phase-contrast method. It allows investigating the broader class of microscopic objects as to their sizes and refractive indices. Moreover, the interference-contrast method is more sensitive because it allows obtaining the maximal contrast of images of phase microscopic objects. But these advantages are neutralized by structural and operation complexity of the interferometers. The method of polarization contrast allows obtaining high quality images with maximal contrast when compared with other techniques, but it can be applied only for anisotropic phase microscopic object, which changes the polarization of light passing through them.

The described classical microscopy methods solve only the problem of phase microscopic objects visualization. The problem of 3D imaging of phase microscopic objects has not even been set within the framework of classical light microscopy. And until the present time electron microscopy was the only method of phase microscopic objects 3D visualization.

Chapter 2

Holography and Holographic Microscopy

2.1 Holography as the method of wave recording and reconstruction

Holography is the method of wave recording and reconstruction which is based on registration of an interference pattern, produced by two coherent waves: the object wave and reference one. The recorded interference pattern is called a hologram. Let us consider the optical arrangement for recording holograms of transparent objects. The optical layouts for recording and reconstruction of holograms are presented in Fig. 4.

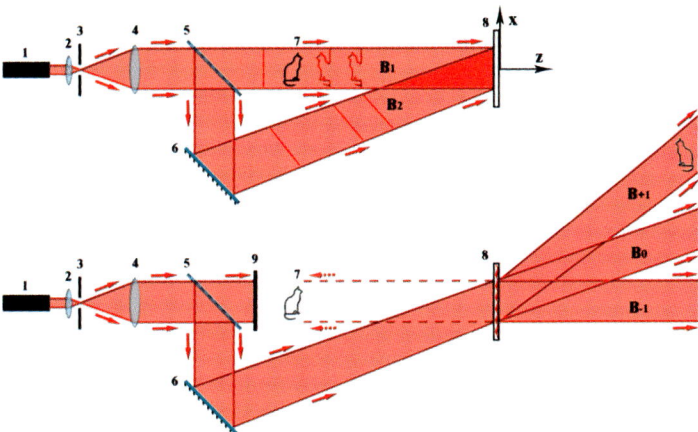

Fig.4. The optical layout for recording (at the top) and reconstructing (at the bottom) holograms of transparent objects. *1*- laser; *2,4* – lenses; *3*-diafragm; *5*- semitransparent mirror; *6*- mirror; *7*- object; *8*-hologram; *9*- opaque screen.

The beam of coherent light (Fig.4a) from the laser 1 is broadened by the lenses 2 and 4.The semi-transparent mirror 5 divides the broader beam in the two beams: the object beam and the reference beam. The reference beam B_2 is directed to the mirror 6 and is reflected in the direction of the holographic plate 8.The object beam B_1 passes through the object under study 7 and is also directed to the holographic plate 8. The beams are coherent. They interfere and their intereference pattern (hologram) is recorded on the holographic plate 8.

Let us express this mathematically. The hologram is produced by the interference of the object wave B_1 transmitted through the transparent object 7 and the reference wave B_2. Suppose the waves have identical polarizations. Suppose that $A_1(x, y)$ is the complex amplitude of the object wave B_1 and $A_2(x, y)$ is the complex amplitude of the reference wave B_2 in the plane of the photoplate 8:

$$A_1(x, y) = a_1(x, y)\exp[-i\phi_1(x, y)]$$
$$A_2(x, y) = a_2(x, y)\exp[-i\phi_2(x, y)]$$
(31)

where $a_1(x, y)$ and $\phi_1(x, y)$ are the amplitude and the phase of the object wave; $a_2(x, y)$ and $\phi_2(x, y)$ are the amplitude and the phase of the reference wave.

The complex amplitude of the resulting wave in the plane of the hologram 8 is equal:

$$A(x, y) = A_1(x, y) + A_2(x, y)$$
(32)

and, consequently, the intensity is equal to:

$$I(x, y) = A(x, y)A*(x, y),$$
(33)

where $A*(x, y)$ is the complex conjugate value.

Suppose that the wave transmitted through the object is directed on the holographic photoplate at the right angle, and the object change only the phase along the x axis:

$$a_1(x, y) = a_1, \phi_1(x, y) = \phi_1(x)$$
(34)

Suppose that the reference wave is directed on the photoplate at the angle θ (see fig.4):

$$a_2(x, y) = a_2, \; \phi_2(x, y) = \frac{2\pi}{\lambda} x \cdot \sin\theta \;, \qquad (35),$$

where λ is the wavelength of the used light.

Then the intensity in the interference pattern, produced by the waves, interacting in the plane of the photoplate 8, is determined by the main interference equation (11):

$$I(x) = a_1^2 + a_2^2 + 2a_1a_2 \cos\left[\frac{2\pi}{\lambda} \cdot x \sin\theta - \phi_1(x)\right], \qquad (36)$$

where a_1^2 is the intensity of the object wave, a_2^2 is the intensity of the reference wave. From this formula one can see that the interference pattern is the alternation of dark and bright fringes. The positions of the dark and bright interference fringes depend on the phase $\phi_1(x)$ of the object wave as follows:

$$x_{\max} = \frac{[2m\pi + \phi_1(x)] \cdot \lambda}{2\pi \sin\theta} \qquad (37)$$

$$x_{\min} = \frac{[(2m+1)\pi + \phi_1(x)] \cdot \lambda}{2\pi \sin\theta}, \qquad (38)$$

where $|m| = 0, 1, 2....$

So, the information on the phase of the object wave is coded in the arrangement of the interference fringes recorded on the holographic plate.

Information on the amplitudes of the waves is coded in contrast v of the interference fringes:

$$v = \frac{I_{\max} - I_{\min}}{I_{max} + I_{\min}}, \qquad (39)$$

where $I_{\max}, I_{\min}$ are the maximal and minimal values of interference fringes intensity.

Using Eq. (36) we can get:

$$v = \frac{2a_1a_2}{a_1^2 + a_2^2}. \qquad (40)$$

So, the amplitude of the object wave a_1 is uniquely determined if the amplitude of the reference wave a_2 is constant.

Thus, the recorded pattern of interference of the object and the reference waves (a hologram) contains information on the phase and amplitude of the object wave.

The methods of recording holograms can be different. The hologram can be recorded as variations of the absorption coefficient (amplitude holograms), or as variations of the refractive index (phase holograms). The hologram can be recorded on a surface (two-dimensional recording) or in a volume (holograms in the three-dimensional medium).

The object considered above is very simple (it changes the phase only in one direction). But the results of the consideration are suitable for very complicated objects. As the result, the recorded hologram will be also very complicated.

Let us consider the reconstruction process (Fig.4b). The hologram 8 (the developed holographic plate) is illuminated by the same reference wave which was used in the process of the hologram recording, when the object beam is screened by the opaque screen 9.Due to the process of light diffraction on the interference structure of the hologram, we obtain the three beam : the direct beam B_0 and the two diffracted beams B_{-1} и B_{+1}`.The two diffracted beam are responsible for the reconstruction of the 3D image of the object .

Let us describe mathematically the process of the reconstruction. We are going to consider the case, when the distribution of the intensity in the interference pattern is recorded as the variations of the absorption coefficient. Such situation occurs under recording holograms on the industrial holographic plates. In this case the amplitude of the wave transmitted through the hologram depends on the amplitude transmission of the hologram τ_a. In the case of the linear dependence of the amplitude transmission of the hologram from the light intensity

$$\tau_a = \kappa I(x, y) , \qquad (41)$$

where κ is constant and $I(x, y)$ is the intensity distribution in the recorded interference pattern (hologram).So, if the hologram 8 is illuminated by the wave, which is identical to the reference wave, then the complex amplitude of the transmitted wave is equal to:

$$A^{'}(x, y) = \tau_a A_2 = \kappa I(x, y) A_2 . \qquad (42)$$

Substituting the Eqs. (31) and (36) into Eq. (35), we obtain:

$$A'(x) = k_1 a_2 \exp\left[-i\frac{2\pi}{\lambda}x\sin\theta\right] + k_2 a_1 \exp\left[-i\phi_1(x)\right] +$$

$$+k_2 a_1 \exp\left[i\phi_1(x) - 2i\frac{2\pi}{\lambda}x\sin\theta\right],$$ (43)

where $k_1 = \kappa\left(a_1^2 + a_2^2\right)$; $k_2 = \kappa a_2^2$

This equation can be considered as the main holography equation. The first value corresponds to the direct wave B_0 in Fig. 4h; the second value corresponds to the diffracted wave B_{-1}, which forms the virtual image (the reconstructed object wave); and the third value corresponds to the diffracted wave B_{+1}, which forms the real image. As it can be seen from Eq. (43), the phase of the wave B_{+1} differs by the value $-2i\frac{2\pi}{\lambda}x\sin\theta$ from the phase of the original object wave, and that's why the real image is deformed.

Thus the holographic method allows recording and reconstructing the amplitude and the phase of the object wave; the reconstructed virtual image is the replica of the original object wave. So, holography allows 3D imaging of objects.

2.2 History of holographic microscopy

Holography was proposed by D. Gabor [5] as the method for electron microscope improvement. He proposed to record the hologram by an electron beam and reconstruct it in the optical wavelength range. The main magnification in the method is achieved without application of optics and determined by the ratio $\frac{\lambda_{opt}}{\lambda_{el}}$, where λ_{opt} is the wavelength of the reconstructing wave in the optical range; λ_{el} is de Broglie wavelength of the recording wave. D. Gabor based the proposed method theoretically and demonstrated holographic process of recoding and reconstruction of an image in the optical range. So, it would be very attractive to record a hologram using X-radiation and reconstruct it using the light of optical range. In such a manner it would be possible to obtain result which gives electron microscopy. Because of the absence coherent source of electron and X-radiation, this method has not found the application in microscopy.

But holography has opened up new possibilities for optical microscopy. The inventor of holography D. Gabor introduces an optical scheme in which the interfering waves travel in one direction (axial holograms).In his scheme the interference pattern of the light diffracted on the microscopic object and the light which passes through the microscopic object, is recorded. D. Gabor showed the applicability of this new process of wave front recording by using a mercury discharge lamp and taking collinear object and reference beams. The original in-line technique of D. Gabor produces both virtual and real images on the same axis, thus an observer focusing on one image, always sees it accompanied by the out-of-focus twin image.

Further development in this field was stymied during the next decade because light sources available at that time were not truly coherent. The situation changed in 1960 with the invention of the laser, whose coherent light was ideal for making holograms. In 1962 E.N. Leith and J Upatnieks for the first tome used a laser for hologram recording and proposed the off-axis optic scheme [7].In the off-axis scheme two beams of light interfere: an object beam and a reference beam. The object beam passes through the object, and the reference beam passes by it. The off-axis hologram generates virtual and real images angularly separated from each other and from the direct beam also (Fig.4).

At the beginning of holographic microscopy development the possibility of holographic image magnification without use of optical systems was investigated. The magnification can be obtained by use of different wavelength of light in the stages of recording and reconstruction of holograms, by increase of hologram's size or by use different divergence of light beams of recording and reconstructing waves. The first optical layout of the optical holographic "lensless" (without an objective) microscope was proposed by E.N. Leith and J Upatnieks [8].The magnification in the microscope was reached by use of light beams with different divergence on the stages of recording and reconstruction of holograms. The image with magnification 150x and resolution equal to 5 μm was obtained.

To obtain high magnification, it is necessary to use an objective in the scheme of the microscope.

The most successful application of holography to microscope has begun with the system in which holography is combined with conventional microscopy. The first holographic microscope using an optical microscope in its scheme was described by R.F. van Lighten and G. Ostenberg in 1966 (Fig.5). The image of the microscopic object magnified by the optical microscope was recorded on the hologram. Holograms of brain neural net with the optical resolution about 1 μm were obtained using the microscope [9].

An optical resolution is the main characteristics of a microscope. The optical resolution of classical microscope is determined by two quantities: the wavelength λ of the radiation being used and by the quantity of the numerical aperture of the objective $n \sin u$. In the case when non coherent radiation is used, the minimal resolved element of a microscopic object is determined by the formula [2]:

$$d \geq \frac{0.6\lambda}{n \sin u},$$ (44)

where u is the aperture angle, n is the refractive index of the ambient media. Using immersion liquid, it is possible to increase the quantity of the numerical aperture to 1.35; and the minimal resolvable distance between two elements of the microscopic object under study can be about half of the wavelength of the radiation being used. The coherent laser radiation is used in the holographic microscope to illuminate the microscopic object under study. In this case the optical resolution of the microscope is determined by the formula:

$$d \geq \frac{\lambda}{n \sin u}.$$ (45)

Thus, the minimal resolvable distant between two details of the microscopic object under study is of the order of the wavelength. Decrease of the optical resolution of holographic microscopes is the principle shortcoming of the holographic method.

The next step in the development of holographic microscopy involved increasing the field of vision and the depth of the recorded scene. G. W. Ellis [3] proposed to place the hologram between the objective and eyepiece of the microscope (Fig. 6). The reconstructed image is studied through the eyepiece which can be moved along the depth and all over the field of vision of the recorded scene. The general magnification of the microscope, which is determined either by the objective, or by the eyepiece, is maintained.

The main advantage of the holographic microscope consists in the possibility of a posteriori study of the reconstructed image, to which all the known methods of the optical and digital processing can be applied. That is very important in the case of short-living and moving microscopic objects.

Though the holography allows recording and reconstruction of 3D images of the microscopic object, to obtain contrast images of phase microscopic objects special methods must be applied. For visualization of phase microscopic objects reconstructed from a hologram the F. Zernike's phase-contrast method was used for the first time in these holographic microscopes. As it was shown above, the method is not the

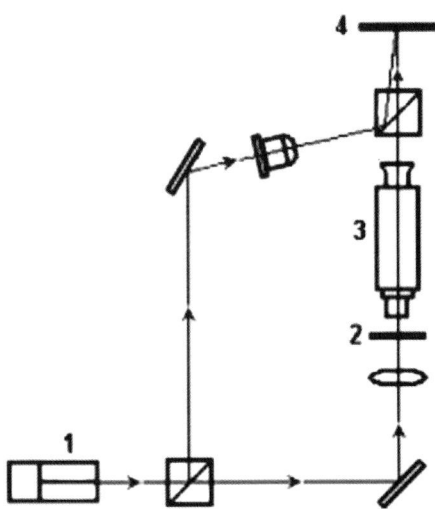

Fig.5. The optical layout of the holographic microscope: *1*- laser; *2* – microscopic object under study; *3* – microscope; *4*-hologram

most effective for obtaining contrast images of phase microscopic objects. The methods interferomertry are more effective for this purpose. Development of holography resulted in the possibility to use the method of interferometry in microscopy, and the holographic analogs of the classical microscopy methods for phase microscopic object visualization have appeared. They are holographic phase-contrast method, the method of holographic interferometry , and the holographic polarization contrast

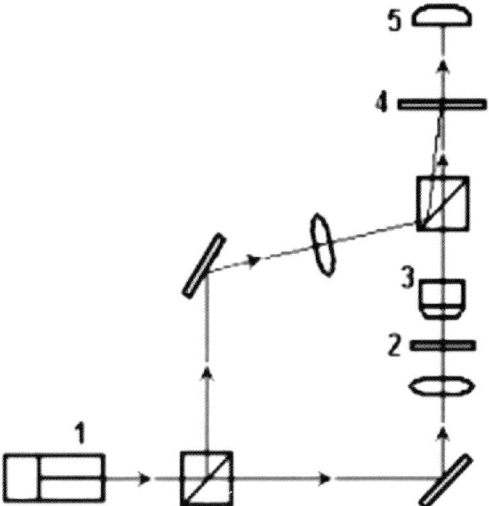

Fig.6. The optical layout of the holographic microscope. *1*-laser; 2-microscopic object; *3*-microobjective; *4* –hologram; *5*- eyepiece.

methods. For the first time phase-contrast images and interferograms in fringes of finite width of native cells were obtained. Application of the holographic methods in microscopy has opened qualitatively new possibilities in microscopy of phase microscopic objects. Now it was made possible not only to visualize phase microscopic objects but also to measure their geometrical parameters using their interferograms. Combining the holographic methods with the methods for digital image processing has made it possible to create the digital holographic interference microscope (DHIM) for real-time 3D imaging of phase microscopic objects and measurement of their geometrical parameters. The instrument integrates holographic interference microscope with digital processing of interferograms. For the first time it has become possible to obtain 3D images of native cells.

Chapter 3

Holographic Methods of Phase Microscopic Object Imaging and the Digital Holographic Interference Microscope

3.1 Holographic phase-contrast method

The classical F. Zernike phase-contrast method was used for obtaining contrast images of phase microscopic objects reconstructed from a hologram in holographic microscopy. The images of the phase microscopic object against dark and bright background were obtained by placing the phase plate into the zeroth order of the diffraction spectrum of the microscopic object. As was shown above, the classical phase-contrast method has some shortcomings and is not the most effective for phase microscopic objects vizualizations. The problem of phase-contrast realizing in holographic microscopy has remained open. The interference-contrast method in an infinitely wide fringe is more efficient for visualization of phase microscopic objects. If the advantages of the method are neutralized in the classical microscopy by the difficulties of obtaining two identical waves and adjustment of the interferometer, the use of holography removes these problems. This is because the wave, which is a copy of the object wave, can be recorded and reconstructed from a hologram. However, because of certain difficulties, the method of holographic interferometry in an infinitely wide fringe has not been used for obtaining images of phase microscopic objects. The difficulty consisted of providing the holographic addition and subtraction of the waves. This problem is easy to solve theoretically. The possibility of subtracting waves to obtain a difference image was already predicted in the early papers of D. Gabor. The phase of the wave recorded and reconstructed from an amplitude hologram (the negative of the initial interference pattern), is shifted by π relative to the phase of the initial

31

wave. Therefore, the object wave and its anti-phased copy reconstructed from the hologram must quench each other when they are observed simultaneously (subtraction of waves). If one of the waves was transmitted through a phase microscopic object, then its bright image must be observed on the dark background. Such holographic subtraction can be realized in the real-time interferometry. On the other hand, when the unperturbed object wave and the object wave perturbed by a microscopic object are recorded on a single hologram, then the addition of waves must be observed, because the phases of both waves will be shifted by π relative to the phase of the initial wave. And a dark image of the microobject must be observed on the bright background. So, the holographic addition of waves can be realized by the method of two-exposure interferometry. However, because of phase increments that arise when the holographic emulsion is processed and the very high sensitivity of the method to vibrations, it was not possible earlier to obtain addition and subtraction of the waves in such an experiment. And the problem of obtaining in-phase and anti-phase waves remained open, like the problem of accomplishing phase contrast in holographic microscopy.

The problem of realizing phase contrast in holographic microscopy was solved in 1985 [20].The proposed method was called the holographic phase-contrast method. The holographic phase-contrast microscope realizing the method was created [22]. For the first time contrast images of native blood cells were obtained using the method. The holographic addition and subtraction of the undisturbed and the disturbed by the microscopic objects waves in an interference fringe is performed for obtaining contrast images of phase microscopic objects. The undisturbed ("empty") wave is the wave which is transmitted through the optical elements of the microscope in the absence of microscopic object under study. The disturbed object wave is the wave transmitted through the optical elements of the microscope when the microscopic object is placed in front of the objective. One of the waves is recorded and reconstructed from the hologram using the reference beam. The phase difference between the waves required for realizing subtraction or addition of the waves is obtained by creating a small angle between the waves, so that the period of the resulting interference pattern significantly exceeds the size of the microscopic object image. When this is done, the conditions for waves to be anti-phase (subtraction of the waves) are automatically created within a dark interference fringe, and bright images of the phase microscopic objects are observed on a dark background; while the conditions for the wave to be in phase (addition of the waves) are created within a bright interference fringe, and dark images of the microscopic objects are observed on a bright background. The adjustment of the microscope is implemented by small transversal shifting the hologram.

Let us consider the result of interaction of the undisturbed and disturbed by the microscopic objects waves, one of which is recorded and reconstructed from a hologram. Suppose that intensities of the waves are equal and their polarizations are identical. In accordance with Eqs. (19) and (20), the intensity in the phase microscopic object image $I_{im}(x',y')$ and the intensity of the background I_b in the case of addition of waves (a bright interference fringe) are determined by the expressions:

$$I_{im+}(x',y') = 2I_0\left[1 + \cos\varphi(x,y)\right],$$
$$I_{b+}(x',y') = 4I_0;$$

(46)

and in the case of the subtraction of waves (a dark interference fringe) by the expression:

$$I_{im-}(x',y') = 2I_0\left[1 - \cos\varphi(x,y)\right],$$
$$I_{b-}(x',y') = 0.$$

(47)

where $\varphi(x,y)$ is the phase increment inserted by the microscopic object into the wave transmitted through it; I_0 is the intensity of each of the two waves. As it follows from Eqs. (46) and (47), the minimal intensity in the microscopic object image in the case of the wave addition corresponds to the maximal intensity in the image in the case of the wave subtraction, and vice versa, so the contrast of the images is inverted.

To obtain contrast images of phase microscopic objects one of the two object waves (the "empty" one or the transmitted through the microscopic object) must be recorded on the hologram. Both these variants are used depending on the type of the problem being solved.

The variant of obtaining contrast images of phase microscopic objects recorded and reconstructed from a hologram can be used for study of alive moving microscopic objects or in the cases, when the microscopic objects are accessible for observation during a short period of time. In this case the microscopic object under study is placed in front of the microscope objective, and its hologram is recorded using the reference wave for a short exposure time or even in the pulse mode of laser radiation. The processed hologram is placed into its initial position. The interferogram of the microscopic object is produced by the intereference of the object wave, reconstructed from the hologram, and the "empty" real object wave. Changing the angle between the two waves (by shifting the hologram from its initial position) it is possible to reach the situation when the period of interference fringes on the interferogram of the microscopic object is larger than the dimensions of the microscopic object. In this case dark image of the microscopic object in the bright interference fringe, and bright image of the microscopic object in the dark interference fringe are observed. In this variant a separate hologram is recorded for every microscopic object under study.

In the variant of obtaining contrast images of phase microscopic objects in real time the "empty" object wave (the microscopic object under study is absent in front of the microscope objective)is recorded on the hologram using the reference beam. The microscopic object is placed in front of the objective. The interferogram of the microscopic object is produced by the interference of the reconstructed from the hologram "empty" object wave and the real object wave, transmitted through the microscopic object. It is obvious, that in this case we can observe time-varying changes the microscopic object. In this variant of the method one hologram can be used to study different microscopic objects. But every objective of the microscope needs it own hologram.

The detail of the technical realization of the method will be considered below in the part dealing with the digital holographic interference microscope (DHIM) and its applications.

Experimental studies have confirmed that the proposed method was effective for obtaining contrast images of phase microscopic objects. Untreated blood smears on glass substrates were used as test objects in the experiments. To evaluate the holographic method compared with the classical methods, images of native human blood erythrocytes obtained by light microscopy, phase-contrast microscopy and holographic phase-contrast microscopy are presented (Fig.7). A He-Ne laser with radiation wavelength equal to 0.63 μm was used as a source of light in the holographic interference microscope to obtain the images. The blood smears of different donors were used for obtaining the optical and holographic images, presented in Fig.7. The images illustrate the

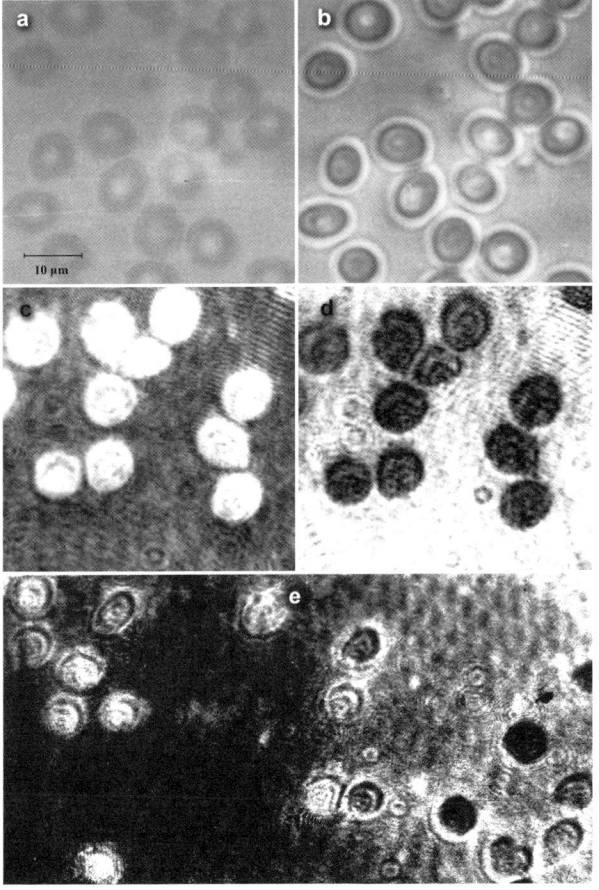

Fig.7. The images of human blood erythrocytes, obtained by the classical and holographic methods (x 400): using the "Lumam" microscope in the white light (a); by F. Zernike method using the phase-contrast attachment in the microscope "Lumam" (b); by the phase-contrast holographic method in a dark interference fringe (c), and in a bright interference fringe (d); and on the boundary of the interference fringes (e). The holographic images are obtained using the holographic interference microscope.

efficiency of the holographic phase-contrast method for obtaining contrast images of phase microscopic objects. The contrast of the images obtained by the holographic subtraction and addition of waves are maximal. But coherent noise caused by diffraction of laser radiation on defects of optical elements and the hologram worsens the quality of the holographic image.

Using the method it is possible to observe contrast images of phase microscopic objects on the dark and bright background simultaneously.

The intensity of images on the interferograms of microscopic objects is modulated by the phase increments, inserted by the microscopic object into the transmitted wave. According to the Eqs. (46) and (47) the phase increment is equal:

$$\varphi(x, y) = \arccos[\pm(1 - \frac{I_{im\mp}(x', y')}{2I_0})] \tag{48}$$

where $I_{im\mp}(x', y')$ is the intensity of the microscopic object image; I_0 is the intensity of each of the two interfering waves; the sign "+" corresponds to the case of wave subtraction; the sign "-" corresponds to the case of wave addition. In accordance with Eq. (46), $I_0 = \frac{I_{b+}}{4}$, where I_{b+} is the background intensity in the case of wave addition. Then

$$\varphi(x, y) = \arccos[\pm(1 - \frac{2I_{im\mp}(x', y')}{I_{b+}})]. \tag{49}$$

So, measuring the intensities in the holographic image of phase microscopic object and the background intensity in the case of holographic addition of waves it is possible to reconstruct the phase relief of the microscopic object under study.

Speaking about interferometry either classical, or holographic, it is necessary to mention its unique property. As it known, the numerical aperture $n \sin u$ of the objective of the microscope is   the larger, the smaller the microscopic object under study. Short-focal-length objectives have small depth resolution, which is inversely proportional to the square of the numerical aperture.

Only a little layer Δz of the microscopic object is seen when it is observed using a microscope (z is the direction of observation). The intensity variations on the interferogram of the phase microscopic object are caused by the phase increment of the transmitted wave through the entire thickness of the microscopic object,

$$\varphi(x, y) = \frac{2\pi}{\lambda} \int_{z_1}^{z_2} \Delta n(x, y, z) dz , \qquad (50)$$

where $\Delta n(x, y, z)$ is the difference of the refractive indices of the microscopic object and the ambient medium. So, interferogram of microscopic object makes it possible to see the whole "phase profile" of the microscopic object under study.

If the microscopic object has a homogeneous refractive-index distribution, i.e. $\Delta n(x, y, z) = const$, we get

$$\varphi(x, y) = \frac{2\pi}{\lambda} t(x, y) \Delta n , \qquad (51)$$

where $t(x, y) = z_2 - z_1$ is the thickness of the microscopic object in the point (x, y). From this we obtain:

$$t(x, y) = \varphi(x, y) \frac{\lambda}{2\pi \Delta n} . \qquad (52)$$

Taking into consideration the expression (48), we get the expression for the thickness of the phase microscopic object:

$$t(x, y) = \frac{\lambda}{2\pi \Delta n} \arccos[\pm(1 - \frac{2 I_{im\mp}(x', y')}{I_{b+}})] \qquad (53)$$

where λ is the wavelength of the radiation being used, Δn is the difference of the refractive indices of the microscopic object and ambient medium, $I_{im\mp}(x', y')$ is the intensity in the corresponding point of the microscopic object image on its interferogram, I_{b+} is the background intensity when the waves are added. The + sign corresponds to the case of subtraction of waves, the – sign corresponds to addition of waves.

Thus, the method of holographic addition and subtraction of waves makes it possible to determine physical thickness of the phase microscopic objects in every point and reconstruct its 3D image. With the development of methods of computer processing of images, the possibility of solving the problem of 3D imaging of phase microscopic objects has appeared.

The following algorithm can be used for computer reconstruction of the 3D image of a phase microscopic object from its interferogram, obtained when the holographic method of phase contrast is used:

The image is transformed into an array of intensities I_{im} at each of its point. The mean background intensity I_{b+} (the mean intensity in a segment chosen by the operator or by the program itself) and the initial intensity $I_0 = I_{b+} / 4$ are found. Next the phase increments are calculated from the formula (49) for each point of the image, and an array is formed corresponding to the phase relief of the wave transmitted through the microscopic object. If the microscopic object has a homogeneous refractive-index distribution and the phase increment is consequently a linear function of the thickness of the microscopic object, the algorithm for constructing a 3D image regards the array of values of the phase increment as an array of values of the third coordinate (the height), taking into account Eq. (52).

For the first time 3D images of native cells (blood erythrocytes) were obtained by the holographic phase-contrast method using the DHIM in 2004 [30].

Human blood erythrocytes are phase microscopic objects with the homogeneous refractive-index distribution. Dry blood smears on glass substrates were used in our experiment, thus Δn is the refractive-index difference of the erythrocytes and air. The refractive index of blood was determined by means of an Abbe refractometer using a He-Ne laser and equal 1.352, the refractive index of air equal 1.These values were used in Eq. (53).Figure 8 shows holographic phase-contrast images of blood erythrocytes (a, b) and their 3D images obtained using the DHIM.

3.2 Holographic interferometry in fringes of finite width

In the holographic method of interferometry in fringes of infinite width (the method of holographic addition and subtraction of waves) the phase increments inserted by the microscopic object into the wave transmitted through it modulate the intensity in the image of a microscopic object, and it becomes visible. As it is shown above, the method of interferometry in fringes of finite width involves the interaction of waves between which there is a certain angle. In this case, instead of homogeneous field, a line field is obtained, with uniform alternation of intensity maxima and minima. So, in the method of interferometry in fringes of finite width the system of interference fringes is modulated by the phase increments inserted by the microscopic object into the wave transmitted through it, and this manifests itself in displacements of the interference fringes. As it follows from Eq. (26), the

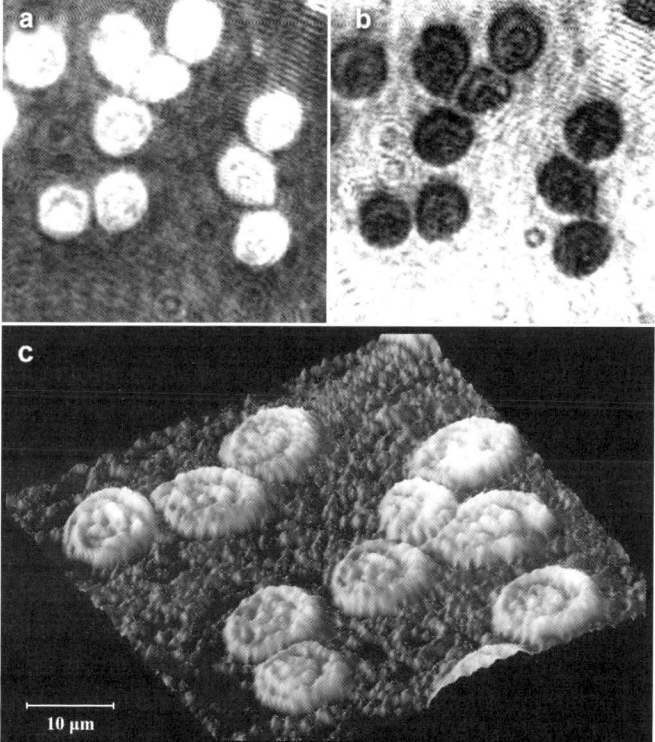

Fig.8. Holographic phase-contrast images of human blood erythrocytes (a, b) and their 3D images obtained using the DHIM.

displacement $h(x', y')$ of the interference fringe in some point of the interferogram is proportional to the phase increment $\varphi(x, y)$ in the corresponding point of the microscopic object, i.e.

$$h(x', y') = \frac{T \varphi(x, y)}{2\pi},\qquad(54)$$

where T is the period of the system of the background interference fringes.

The main difference of the holographic method of interferometry in fringes of finite width from the classical one consists in the following. The system of the background interference fringes in the classical interferometer is produced by the interference of two waves transmitted simultaneously along two paths and re-joined (Fig. 2). The system of background interference fringes in the holographic interferometer is produced by two waves transmitted along the same path but in different instants of time. One of these waves is recorded and reconstructed from a hologram. The holographic interferometer is a single arm interferometer as opposite to the classical interferometer which is a two-arm interferometer. Because the recorded and reconstructed from the hologram wave is a copy of the initial object wave, the problem of obtaining the two identical object waves, which is the main problem of classical interferometry, is completely removed. In comparison with the real-time classical interferometry, holographic interferometry is possible in two variants: real time and two-exposure interferometry. In the real time interferometry only one object wave is recorded on the hologram using the reference wave. The interference pattern (an interferogram) is the result of interference of two object waves: the real wave and reconstructed from the hologram. If the "empty" object wave is recorded on the hologram, then real time interferometry allows obtaining real-time interferograms of the microscopic objects. This method allows observing the time history of the microscopic object under study. In the two-exposure method the two object waves are recorded on a hologram. This method can be used when moving or short living microscopic objects are studied. Moreover, holographic interferometry is possible with any objective. That made it possible to use holographic interferometry for investigation such small objects as cells, fibers, etc.

The method of holographic interferometry in fringes of finite width presents the great advantage of quantitative measurement of parameters, including the phase increment distribution inserted by transparent specimens. As it was mentioned above, the phase increments inserted by the microscopic object into the wave transmitted through it manifest itself in displacements of the interference fringes from their initial positions in its the image. So, measuring the displacements of the interference fringes on the interferogram of the microscopic object it is possible to determine the phase increments $\varphi(x, y)$ inserted by the microscopic object into the transmitted wave and reconstruct the phase profile of the phase microscopic object:

$$\varphi(x, y) = \frac{2\pi h(x', y')}{T} . \tag{55}$$

If the microobject has a homogeneous refractive-index distribution, then it follows from Eq. (51) that

$$\varphi(x, y) = \frac{2\pi}{\lambda} t(x, y) \Delta n . \tag{56}$$

From Eqs. (55) and (56) we obtain that the thickness $t(x, y)$ of the microscopic object at the point is equal:

$$t(x, y) = \frac{\lambda h(x', y')}{T \Delta n} , \tag{57}$$

where $h(x', y')$ is the displacement of the interference fringe in the corresponding point, T is the period of the system of background interference fringes, λ is the wavelength of the radiation being used, Δn is the difference of refractive indices of the microscopic object and the ambient medium..

Measuring the displacements of the interference fringes from their initial positions on the interferogram of the microscopic object it is possible to determine the thickness of the microscopic object in each its point and reconstruct its 3D image.

The algorithm for the mathematical processing the interferogram obtained by the method of holographic interferometry in fringes of finite width consist of interpreting the displacements of the interference fringes as the phase increments in accordance with Eq. (55) for each point of the image, and an array is formed corresponding to the phase relief of the wave transmitted through the microscopic object. If the microscopic object has a homogeneous refractive-index distribution and the phase increment is consequently a linear function of the thickness of the microscopic object, the algorithm for constructing a 3D image regards the array of values of the displacements as an array of values of the heights of the microscopic object in accordance with Eq. (57), followed by its 3D visualization, as well as the determination of its geometrical parameters.   The problem of this type has no exact mathematical solution. However, an approximation algorithm may be used, provided that the error that it introduces does not exceed the error introduced into the image by the fundamental deficiencies of the physical experiment (coherent noise, speckle structure, etc.). Moreover, the problem of reducing the influence of such noise on the image quality can also be solved by computer processing the interferograms.

To obtain interferograms in fringes of finite width the number of interference fringes on an image of a microscopic object must be enough for determination of its shape. When microscopic objects of sizes a few micrometers are studied, speckle structure and coherent noise originating from different sources such as dust particles, scratches, and defects on and in the optical elements become very essential disturbance which blur the interference pattern. This is the main reason why interferometric techniques are widely used in metrology, whereas only a few applications have been reported in biology [6, 16].

Only use of special system for coherent noise averaging made it possible to obtain interferograms of blood erythrocytes [6]. The obtained interferograms allowed making conclusions about shapes of blood erythrocytes. The use of polarization filtering in the holographic interference microscope allowed to obtain interferogram of blood erythrocytes and native blood cells of size about 2 μm [21].

The experiments have confirmed the efficiency of the method of holographic interferometry in fringes of finite width for study of biological phase microscopic objects. Application of digital methods of interferogram processing has made it possible to obtain 3D images of phase microscopic objects by their interferograms. For the first time 3D images of the phase microscopic objects were obtained using the DHIM when investigating the blood erythrocytes of patients suffering from hemolytic anemia [24].

Figure 9 shows interferograms and 3D images of the individual erythrocytes obtained using the DHIM.

Figure 10 shows a segment of an interferogram of an untreated blood smear in fringes of finite width and the reconstructed 3D image of this segment obtained using the DHIM. It is necessary to notice, that there is blood plasma apart from erythrocytes in the native blood smears. When water vaporizes from the plasma, the substances contained in it (in general, proteins) form space structures which are also phase microscopic objects and are observed on the reconstructed 3D images of blood smears. These elements of the plasma form some fibers in the direction of blood drop smearing along the glass substrate.

3.3 Comparison of the holographic methods

The carried out theoretical and experimental comparison of the holographic methods using the DHIM have shown that the method of holographic addition and subtraction of waves and the method of holographic interferometry in fringes of finite width combined with the methods of computer image processing can be used for 3D imaging of phase microscopic objects [32]. 3D images of phase microscopic objects not only give clear qualitative information concerning the shapes of these objects, but they also make it possible to carry out various quantitative measurements.

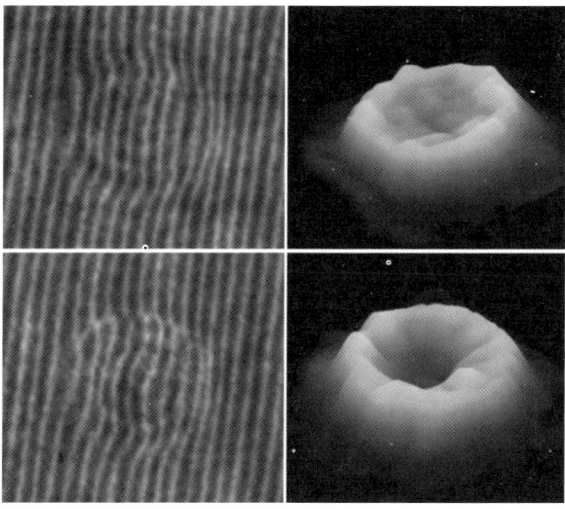

Fig.9. Interferograms of individual erythrocytes (at the left) and their reconstructed 3D images (at the right) obtained using the DHIM.

To compare the methods, interferograms of the same microscopic object obtained by the holographic phase-contrast method and by the method of holographic interferometry in fringes of finite width are presented in Fig. 11.The high level of correspondence of the images obtained by the methods (Fig.11d and 11e) makes it possible to conclude that the computer-processing methods that were used adequately reflect the experimental information and are quite suitable for solving such general problems.

Each method has its own advantages and disadvantages for solving problem of 3D visualization of phase microscopic objects.

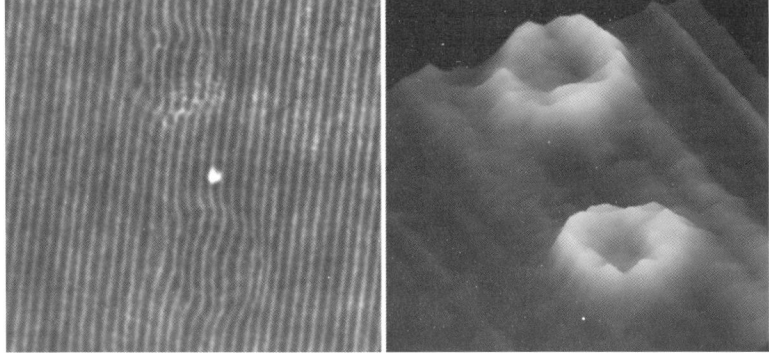

Fig. 10. The segment of an interferogram of an untreated blood smear (at the left) and its reconstructed 3D image (at the right) obtained using the DHIM.

When the method of addition and subtraction in an interference fringe is used, fairy simple exact processing algorithms can be used for computer processing of the interferograms. However, since absolute intensities are used in this method, it is necessary for the transfer function of the image-recording unit to be linear in order to obtain adequate 3D visualization.At the stage of recording the interferogram, this method imposes very high requirements on the vibration stability of the interferometer. Therefore the hologram must be developed in situ, or the microadjusting system is needed.

The problem of 3D visualization of phase microscopic objects when the method of interferometry in fringes of finite width is used has no exact mathematical solution. However, it is allowable to use approximate processing algorithms, provided that the error that they introduce does not exceed the error introduced by the fundamental drawback of the physical experiment (speckle structure, coherent noise). Moreover, computer processing makes it possible to use supplementary algorithms to reduce the effect of these drawbacks (filtering of the speckle defects, averaging of the coherent noise). There are no high requirements on the vibration stability of the apparatus at the interferogram- recoding stage, and the interferometer can operate without any special vibration-proofing measures in such a regime. This makes it possible to use it not only in a specialized optical laboratory, but in any medical or biology ones.

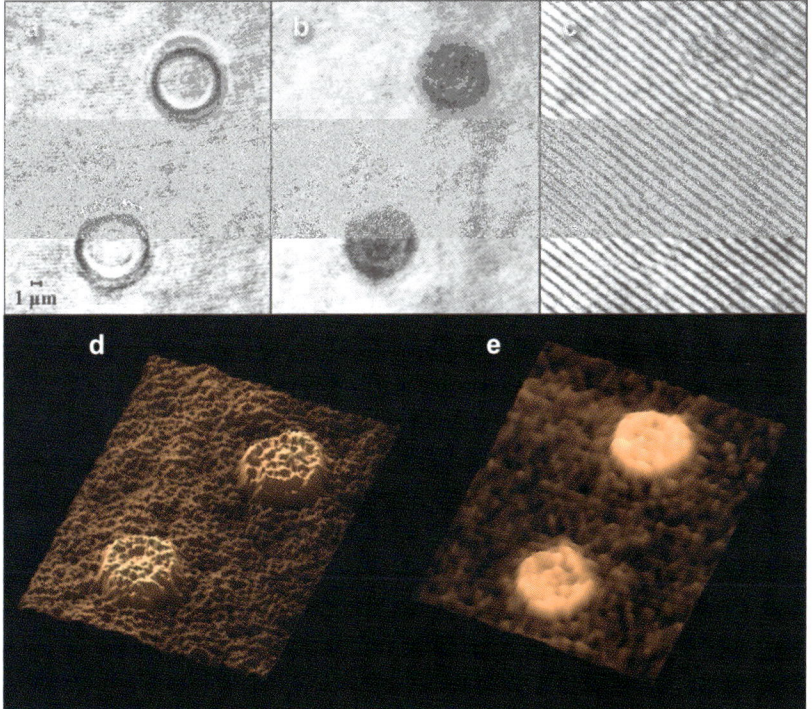

Fig.11. Comparison of the holographic methods for 3D imaging phase microscopic objects. The photograph of blood erythrocytes (a); the holographic phase-contrast image of the erythrocytes in the case of wave addition (b); the interference-contrast image of the erythrocytes (c); 3D images reconstructed from the corresponding interferograms (d and e) .

3.4 Sensitivity of the holographic methods

The sensitivity of the methods is characterized by the minimal measurable value of phase increment φ_{min}. The value which is directly measured in the holographic phase-contrast method is the intensity. As it follows from Eq. (46), when the wave of equal intensities and identical polarizations interact, then the contrast of the interferograms in the case of wave addition is maximal, and the ratio of the intensities of an image I_{im+} and a background I_{b+} in the interference pattern has the form:

$$\frac{I_{im+}}{I_{b+}} = \frac{1}{2}[1 + \cos\varphi(x, y)] = \cos^2[\frac{\varphi(x, y)}{2}]. \qquad (58)$$

Then

$$\varphi(x, y) = 2\arccos\sqrt{\frac{I_{im}}{I_{b+}}}. \qquad (59)$$

So, the sensitivity of the method depends on the possibility of measurement the difference of intensities of the image and the background. For example, if the maximal $\dfrac{I_{im+}}{I_{b+}} = 0.99$, then the minimal measurable phase increment is equal $\varphi_{min} = \dfrac{\pi}{15}$. If the microscopic object has a homogeneous refractive-index distribution, as it follows from equation (52):

$$t(x, y)_{min} = \varphi(x, y)_{min}\frac{\lambda}{2\pi\Delta n}. \qquad (60)$$

Then the minimal measurable thickness is equal to $t_{min} = \dfrac{\lambda}{30\Delta n}$. In the case when a He-Ne laser with the wavelength of radiation $\lambda = 0.63\mu m$ is used for study of dry blood smear, then $\Delta n = 0.352$ (difference of refractive indices of blood and air), and then the minimal detectable thickness of the microscopic object $t_{min} \approx 0.11\lambda \approx 0.07\mu m$.

The value that is directly measured in the method of interferometry in fringes of finite width is the displacement of an interference fringe. The minimal displacement of the interference fringe

$$h_{min} = T\kappa_{min}, \qquad (61)$$

where T is the period of the system of interference fringes, k_{min} is the number of fringes which the interference pattern is displaced on. From the other hand,

$$\varphi_{min} = 2\pi\kappa_{min},\qquad(62)$$

where φ_{min} is the minimal detectable phase increment inserted by the microscopic object. So, if we want to measure the minimal phase increment equals to $\dfrac{\pi}{15}$, then this corresponds to the $\kappa_{min} = \dfrac{1}{30}$ and

$$h_{min} = \frac{T}{30}.$$

In the case of visual detection of the interfernce fringe position, it is considered that $\kappa_{min} = \dfrac{1}{10}$, that corresponds to $\varphi_{min} = \dfrac{\pi}{5}$ [13]. So, application of the computer methods of image processing has made it possible to improve the sensitivity of the holographic interferometry method.

3.5 The holographic polarization-contrast method

The microscopic object under study, as usual, is situated on a glass substrate with other noise microscopic objects (irregularity of the glass substrate, particles of dust, defects, scratches, etc) the sizes of which are comparable with the size of the microscopic object. Diffraction of the coherent radiation on such noise microscopic objects results in coherent noise, which worsens the quality of images and interferograms. The influence of the coherent noise on the image quality is the more essential, the smaller the microscopic object under study. The problem can't be solved using the phase-contrast and interference-contrast holographic methods of microscopic object visualization because these noise microscopic objects are also visualized (for example, the blood plasma in blood smears in Fig.10). This problem can be solved by using the polarization –contrast method when phase microscopic object that possesses the anisotropy property are studied. Polarization-contrast images can be obtained using a polarizer in front of the microscopic object and an analizer (another polarizer) behind the micrbjective in the microscope. As usual, noise microscopic objects do not change the state of polarization of the light transmitted through them. When the polarizers are crossed, bright images of the anisotropic microscopic objects on dark background are observed, and the noise microscopic objects are invisible.

The state of polarization of the reference beam can play the role of the analyzer in the case of hologram recording by the holographic microscope. The holograms of polarization-contrast images of anisotropic microscopic objects can be obtained when linear polarizations of the object and reference beams are orthogonal [18, 19].

As it was shown above, the two waves (ordinary and extraordinary) with mutually orthogonal polarization propagating in one direction inside the anisotropic microscopic object gain phase difference due to the difference of the refractive indices. As it follows from Eqs. (28) and (29), the phase difference $\Delta\varphi$ is equal:

$$\Delta\varphi = \frac{2\pi}{\lambda} t(n_e - n_0), \qquad (63)$$

where n_0 and n_e are the refractive indices for polarizations perpendicular (ordinary) and parallel (extraordinary) to the axis of anisotropy respectively, λ is the wavelength of the light being used, t is the thickness of the microscopic object. The intensity of the image of the anisotropic microscopic object reconstructed from the polarization-contrast hologram depends on the phase difference $\Delta\varphi$. If the microscopic object has different thickness in different point, i.e. $t = t(x, y)$, then the intensity of the image is modulated by the thickness of the microscopic object. This allows one to reconstruct its 3D image. Figure 12 shows the holographic polarization-contrast image of the biological microscopic object (a) and its 3D image (b), obtained using the DHIM.

Figure 13 shows 3D image of the salt microcrystal reconstructed from the polarization-contrast holographic image, obtained using the DHIM.

It is seen that polarization- contrast method improves quality of images in comparison with the phase- and interference- contrast methods due to removing coherent noise and filtering all others microscopic objects.

Contrast and resolution of images of anisotropic microscopic objects obtained by the holographic polarization-contrast method is maximal.

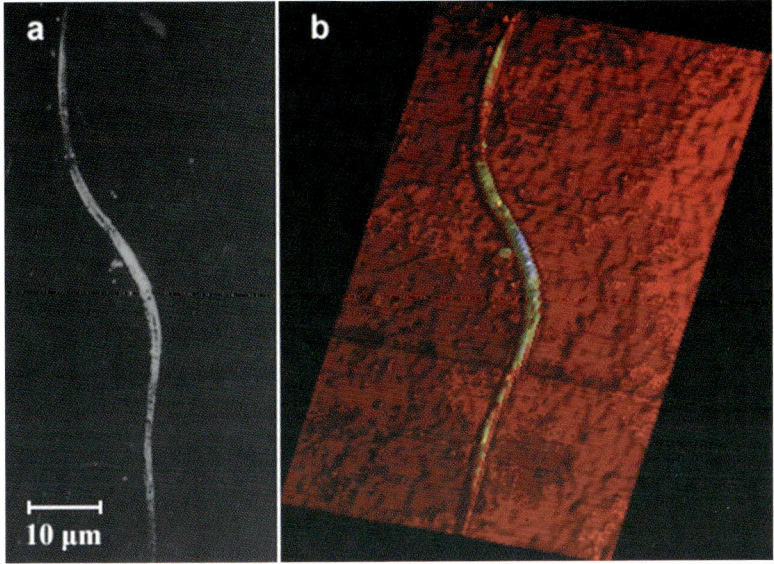

Fig. 12. The polarization-contrast image of the biological microscopic object (a) and its 3D image (b).

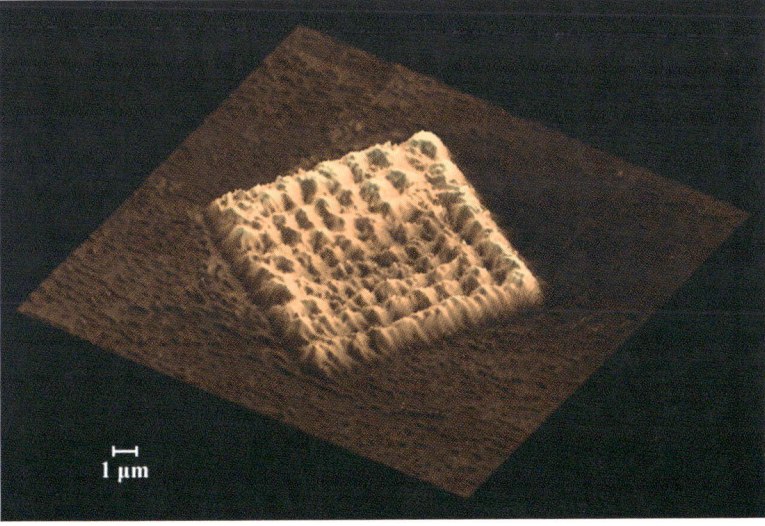

Fig.13. 3D image of the salt microcrystal reconstructed from the holographic polarization-contrast image.

3.6 The digital holographic interference microscope

The first digital holographic interference microscope (DHIM), which allows the real time 3D imaging of phase microscopic object and the quantitative measurements of their parameters, has been created in the Laboratory of Holography, Kharkov National University. The first 3D images of the native cells (human blood erythrocytes) were obtained using microscope in 1998 [24]. The optical part of the microscope and the holographic methods of phase microscopic objects visualization were worked out in 1983-1987 [18-23]. Application of a computer and digital methods of image processing made it possible 3D visualization of phase microscopic objects [24-40].

The DHIM is an easy-to-operate device that allows one to realize all holographic methods of phase microscopic objects investigation: the holographic phase-contrast method (holographic subtraction and addition of waves in an interference fringe), the method of interferometry in fringes of finite width and the polarization-contrast method. The simple and one-axis construction of the DHIM makes it possible to use it as a usual optical microscope, a holographic microscope and a polarization microscope.

The DHIM consists of three main units: holographic interference microscope (holographic microinterferometer), digital video camera and computer. The interferograms and images of the microscopic objects under study obtained using the holographic microinterferometer are recorded by the digital camera. The digital interferograms and images are computer processed using the mathematical algorithms that makes it possible to reconstruct the 3D images of microscopic objects and to measure their geometrical parameters.

The optical layout of DHIM is shown in Fig.14.

A He-Ne laser that emits at 0.63 μm is used as a coherent radiation source. The radiation of laser *1* is divided into two beams by semitransparent mirror *5*: the object beam and the reference beam. The object beam passes through microscope objective *11* and is directed to holographic plate (hologram) *12*.The reference beam passes through collimator *9* and also is directed to hologram *12*. Mirrors *3* and *4* are introduced into the system to rotate the rays. The reference beam plays an auxiliary role in the system. It is needed for recording and reconstructing the object wave from the hologram.

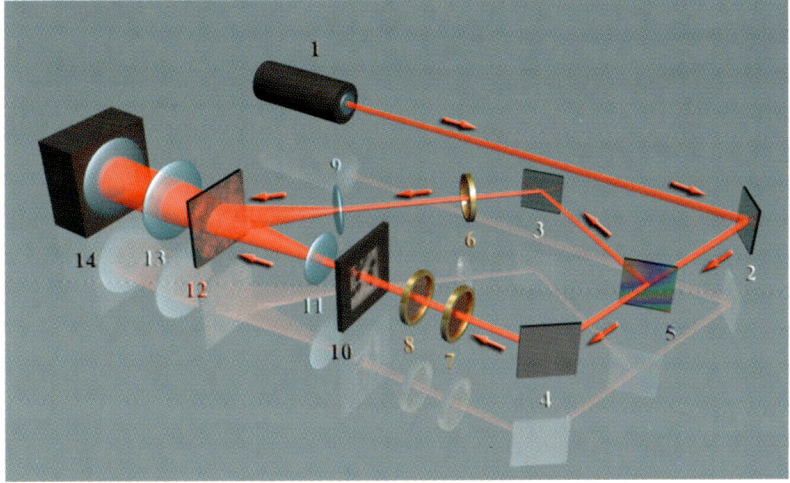

Fig.14. Optical layout of the digital holographic interference microscope. *1*-laser; *2-4* – mirrors; *5*- semitransparent mirror; *6,7* and *8* - polarisers; *9*- collimator; *10*- test specimen; *11*-objective; *12*-hologram; *13*-eyepiece; *14* – image-recording unit.

A hologram of the "empty" object wave is recorded on holographic plate *12* in the absence of the specimen. The developed and fixed hologram is returned to its original position and serves as a permanent optical element of the microscope. The holographic microscope with the hologram becomes a holographic microinterferometer. The hologram, on which the wave, passed through all optical elements of the microscope is recorded, replaces the second shoulder of the classical interferometer. Because the wave reconstructed from the hologram is a replica of the real object wave, the problem of obtaining two identical wavefronts is completely removed. When the wave reconstructed from the hologram by means of the reference beam and the real object wave propagate simultaneously, their interference pattern can be observed by means of eyepiece *13*. The interferogram of the object in classical interferometer is obtained due to interference of two waves passing simultaneously different ways. In the holographic interferometer two waves passed the same way, but in different instants of time. Handling the interferometer is rather simple. It is implemented by small transversal shifting the hologram *12* from its initial position using the micrometer screws of the special holder. As a result a certain angle θ is created between the interfering waves. This angle determines the period of interference fringes:

$$T = \frac{\lambda}{2\sin(\theta/2)},$$

where λ is the wavelength of the laser radiation being used. If $\theta \approx 0$, the holographic microinterferometer is adjusted on the infinitely broad fringe. If $\theta > 0$, then the holographic interferometer is adjusted on fringes of finite width. The period of the interference fringes is the smaller, the larger the angle. This allows obtaining the optimal quantity of interference fringes on images of microscopic objects of different sizes. Hologram *12* is placed between objective *11* and eyepiece *13*.Since an unfocused image is recorded on the hologram, such a placement of the hologram makes it possible to increase the field of view of the microscope because of the possibility of displacing the eyepiece in a plane orthogonal to the observation direction, and to carry out additional focusing over the depth of the observed scene. When the test specimen *10* is placed in front of the microscope objective *11*, its interferogram is observed.The hologram recorded for the given objective can be used permanently when working with the given objective. If we work with different objectives, then a set of hologram for every objective is needed. The holographic interference microscope also can operate as usual optical microscope with laser illumination and as a holographic microscope. The main shortcoming of holographic microscopy is the presence of coherent noise originating from different sources such as dust particles, scratches, and defects on and in optical elements and stray radiation. The coherent noise is the more essential, the smaller the microscopic object under study. To obtain high quality of images, the quality of interferograms must be maximal. As it well known, the quality of holograms and interferograms are determined by such a quantity as contrast:

$$V = \frac{I_{max} - I_{min}}{I_{max} + I_{min}}, \tag{64}$$

where I_{max} and I_{min} are, correspondingly, maximal and minimal intensities in the interference pattern. As it follows from (11) and (12):

$$I_{max} = I_1 + I_2 + 2\sqrt{I_1 I_2}\cos\beta$$
$$I_{min} = I_1 + I_2 - 2\sqrt{I_1 I_2}\cos\beta \tag{65}$$

And from here:

$$V = \frac{2\sqrt{I_1 I_2}}{I_1 + I_2}\cos\beta = \frac{2\sqrt{b}}{b+1}\cos\beta. \qquad (66)$$

Here $b = \frac{a_1^2}{a_2^2} = \frac{I_1}{I_2}$ is the ratio of the intensities of the interfering waves;

β is the angle between the directions of linear polarizations of the waves. The contrast of the interference pattern is maximal and equal to 1, when $b = 1, \beta = 0$, i.e. if the intensities and polarizations of the waves are equal. To meet the conditions, polarizes *6, 7,* and *8* are introduced into the system to equalize the intensities and polarizations of the interacting waves. The state of polarization of the wave reconstructed from the hologram is determined by the state of polarization of the reference beam. Polarizes *6* and *8* situated in the reference and object beams set identical polarizations of the interacting waves. The intensity of the light transmitted through two linear polarizers *7* and *8,* is determined by the relation:

$$I = I_0 \cos^2\alpha, \qquad (67)$$

where α is the angle between the transmittance planes of the polarizers, I_0 is the intensity of the real object wave. The intensity of the wave reconstructed from the hologram is smaller than the intensity of the real object wave. Rotating the transmittance plane of polarizer *7* relative to the fixed transmittance plane of polarizer *8,* it is possible to equalize the intensity of the interacting waves, decreasing the intensity of the real object wave. So, the use of such polarization filtering in the holographic interference microscope makes it possible to reduce the influence of the coherent noise and to obtain interferograms of the microscopic objects of size a few micrometers with the quantity of interference fringes on their images that is optimal for their 3D imaging. Moreover, with the polarizers, the microscope can operate as polarization microscope, and the method of polarization contrast can be used for anisotropic microscopic objects contrast imaging. Interferograms and images of the microscopic objects being investigated are recorded by means of image-recording unit *14,* containing digital camera. A 40 x 0.65objective and 10X eyepiece are used in the microscope. The holograms were recorded on PFG-03 plates. Developer GP-2 was used for processing the holograms. Figures 15 and 16 show the images of native blood cells obtained using the DHIM.

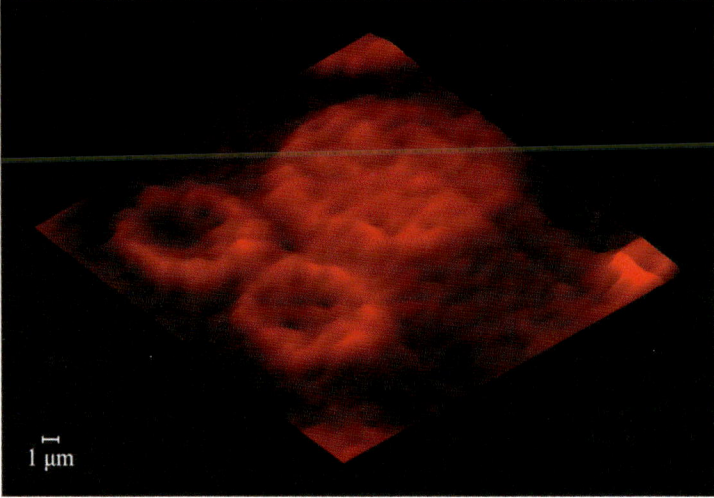

Fig.15. The fragment of native blood smear with a leukocyte and two erythrocytes obtained using the DHIM.

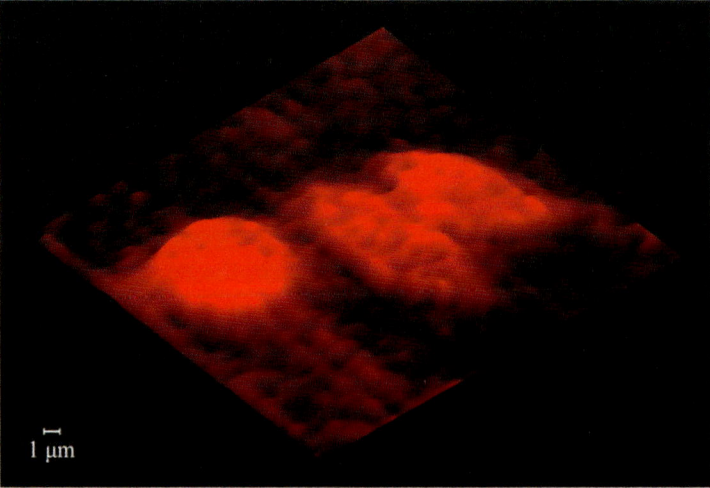

Fig.16.The fragment of the rabbit blood smear with pathological erythrocytes obtained using the DHIM.

3.7 Digital holographic microscopy

Research activity in the past years has been directed at improving numerical methods of hologram processing. This results in appearing a lot of digital holographic microscopes which are based on a digital holography principle [10, 14]. Digital holography, i.e. digital recording and numerical reconstruction of holograms, offer new perspectives in imaging, because numerical proceeding of a complex wave front allows one to compute simultaneously the intensity and phase distribution of the wave. In microscopy digital holography has made it possible to focus numerically on different object planes without using any optomechanicalmovement. Different lens aberrations can be corrected by a numerical procedure. The first digital holographic microscope (DHM) was presented in 2005 [10]. The absolute phase distribution images of living neurons in a culture were obtained by use of a DHM. In opposite to the DHIM which is basically a holographic interferometer, the optical part of the DHM is a classical Mach-Zehnder interferometer. The interferograms of microscopic objects are numerically proceeded in the DHIM, whereas holograms of the microscopic object are used in the DHM for 3D imaging.

Chapter 4

Application of the Digital Holographic Interference Microscope for Blood Cells Study

In this book we present our pioneer results of DHIM application for experimental study of the 3D morphology of biological and technical phase microscopic objects. They are human blood erythrocytes and thin films.

4.1 DHIM study of 3D morphology of blood erythrocytes

The study of the interaction of a living organism with the environment at a cellular level is of great scientific interest. A cell is a unit of a living organism that reflects and determines the status and functioning of the biological system as a whole. A number of papers have appeared in recent years in which it was established that the main targets of the action of various physical and chemical influences of the environment, as well as of internal pathologies and even of the mental states are cellular plasmatic membranes. The variation of the elastic properties of the cellular membranes under the action of various effects must unavoidably show up in shape modifications of the cells. The blood cells are of special interest. Blood unites the operation and functioning of all organs of the living organism. The red blood cells (erythrocytes) are the cells whose main function is the oxygen transport from lungs to tissues and the carbon dioxide transport to lungs. In addition, erythrocytes take part in the processes related to maintaining of the homeostasis at the organism level. Moreover, an erythrocyte is available for observation as a separate cell of a living organism. An erythrocyte is a cell without nucleus, and the content of protein hemoglobin in it is 98

%. For the optimal functioning, the functional activity of hemoglobin must be maintained, and the optimal 3D shape of erythrocyte must be realized. The erythrocyte shape must correspond to maximum surface at the given volume and must ensure deformations providing the erythrocyte motion along thin capillaries. These conditions are satisfied for a biconcave disk shape, which is considered to be the medical norm. The first image of erythrocytes as biconcave disks was obtained using electron microscopy [17]. Then, this result has been confirmed. It is commonly accepted that the fraction of the biconcave erythrocytes in the blood of healthy patients ranges from 60 to 97%.It is knows that in the case of hematological diseases the modifications of erythrocyte shapes are observed. The results of electron microscopy show that, in addition to hematological diseases, the diseases of various genesis can be the reason for modifications of the 3D shape of erythrocytes [11]. This made it possible to formulate the concept of the erythrocyte pathomorphosis. However, this concept was under question, since the morphological modifications observed using electron microscopy can be caused by various effects related to the sample preparation.

Red blood cells, like other cells of a living organism, are phase microscopic objects. Therefore, the problem of studying the morphology of untreated erythrocytes for purposes of medical diagnosis involves the problem of 3D visualization of the phase microscopic objects and measuring their morphological parameters.

Human erythrocytes are microscopic objects with a size 7-8 μm and a maximum thickness of about 2 μm. Each one is a microscopic object with a low optical density and homogeneous refractive-index distribution, transparent for radiation of a He-Ne laser. The optical homogeneity of the erythrocytes makes it possible to compute the thickness of the erythrocytes in different points and reconstruct their 3D images using their interferograms. It is known that erythrocytes do not change their shape after drying. That makes it possible to use dry untreated blood smears on glass substrates, which are normally used in medical practice, to study 3D morphology of blood erythrocytes. Then, Δn in Eq. (57) is the difference of the refractive indices of the erythrocytes and air. The refractive index of blood was measured by means of Abbe refractometer using a He-Ne laser and equals 1.352. This value was taken as refractive index of the erythrocytes. The refractive index of air equals 1. So, Δn is equal to 0.352.

This value was used in Eq. (57) under computer processing the interferograms of erythrocytes for their 3D imaging and measuring their morphological parameters.

Figure 17 shows the reconstructed 3D image of a fragment of the untreated blood smear of a healthy donor obtained using the DHIM. It is seen that biconcave erythrocytes dominate in the blood smear. One can see the direction of blood smearing.

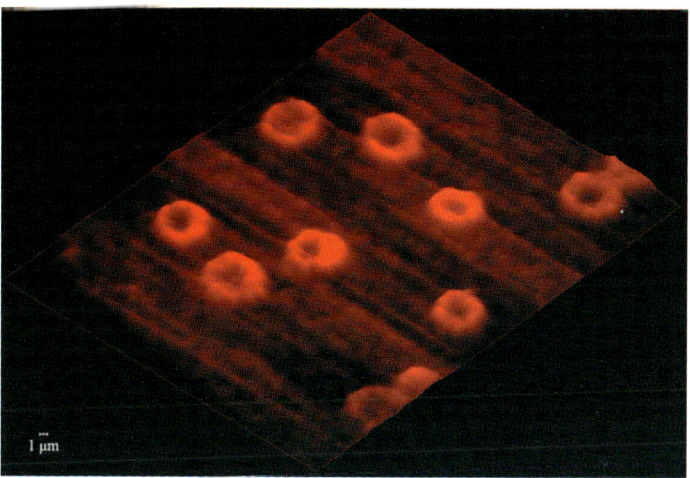

Fig. 17. 3D image of the fragment of a human blood smear obtained using the DHIM

To compare the method of holographic interferometry and the method of electron microscopy, a fragment of a blood smear obtained by the electron microscopy method, is presented in Fig.18 [12].

It is necessary to notice that Fig. 17 shows the image of untreated, native erythrocytes. The method of holographic interferometry is not destructive for specimens; the specimen is not damaged in the DHIM, and it can be investigated once more and by any other optical methods. Figure18 shows the image of blood erythrocytes after the multistage preliminary preparation (dehydration, metallization) which is necessary for obtaining images by the electron microscopy method. The method of electron microscopy is destructive for biological specimens; the specimen can not be used once more.

In our experiments we have determined three main morphological types of erythrocytes: biconcave disk, flat disk, and spherocytes [12, 33, 34]. To characterize the morphological type of erythrocytes, a sphericity coefficient k is introduced as a ratio of the erythrocyte thickness at the center d_c to the thickness at half of its radius d_r (Fig. 19):

$$k = \frac{d_c}{d_r}. \tag{67}$$

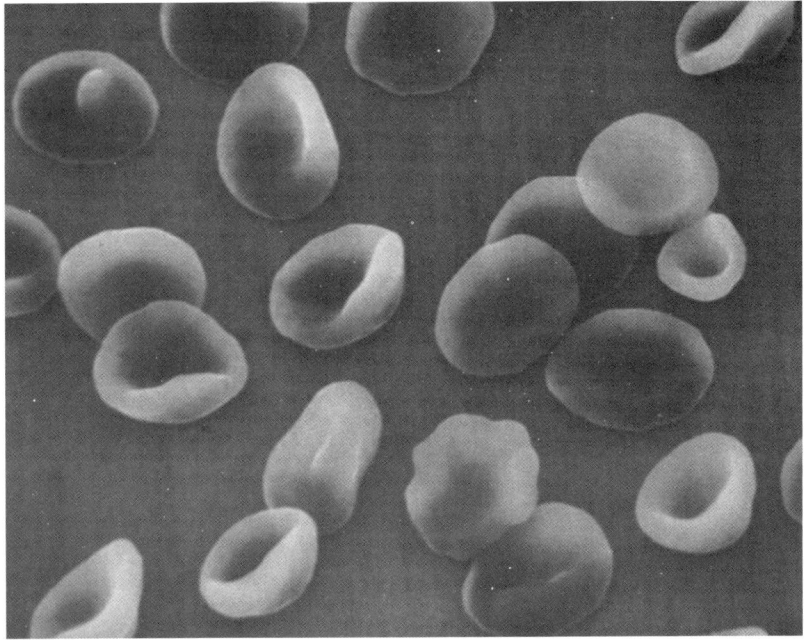

Fig.18. 3D image of blood erythrocytes obtained using the electron microscope.

For the biconcave erythrocytes, flat disks, and spherocytes, the sphericity coefficient is less than unity, about unity, and greater than unity, respectively. The sphericity coefficient is measured upon the computer processing of the DHIM erythrocyte interferograms.

Figure 20 shows 3D images of the three main morphological types of erythrocytes and their sphericity coefficients.

The biconcave erythrocytes correspond to the medical norm; the functional possibilities of such erythrocytes are maximal (Fig.20a).

Functional possibilities of flat (Fig.20b) and spherical erythrocytes (Fig.20c) are reduced.

These main morphological shapes of erythrocytes and their different variants were detected in the native blood smears of donors and patients

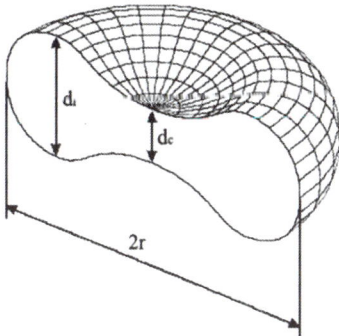

Fig.19. The sphericity coefficient k determination. d_c is the thickness in the center; d_r is the thickness at a half of the radius r.

with different diseases; but contain of these types of erythrocytes was quite different.

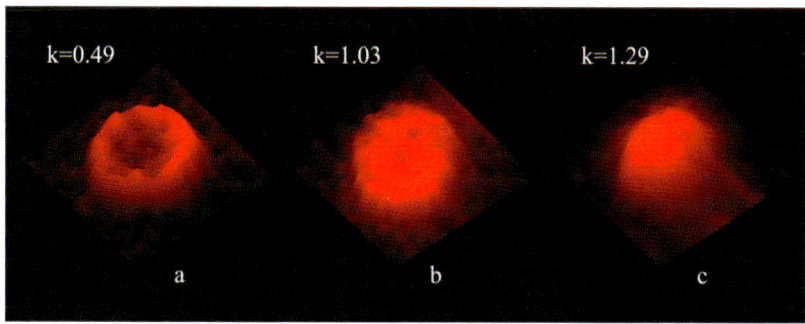

Fig.20. Three main morphological types of blood erythrocytes: biconcave disk (a); flat disk (b); spherocyte (c), and their sphericity coefficients. 3D images are obtained using the DHIM.

Figure 21 shows the fragments of the native smears of blood of a healthy man aged 56 (a) and a diseased woman (cavernous hemangioma) aged 23 (b). The mean sphericity coefficients of the fragments are presented. The mean sphericity coefficient k_m is determined as arithmetical mean of the spericity coefficients of the separate erythrocytes of the fragment.It is seen that the biconcave erythrocytes dominate in the blood smear of the healthy donor (Fig.21a). The mean sphericity coefficient of his smear is equal to 0.35.The fragment of the blood smear of the woman (Fig.21b) contains flat disk erythrocytes and a spherocyte. The mean sphericity coefficient of the fragment of her blood smear is equal to 1.12. The difference of blood smears of the healthy donor and the diseased woman is obvious.

Patients with different diseases were examined in our experiments. The majority of the patients had the morphological modifications of the erythrocytes, the tendency to the increase of the mean sphericity coefficient of blood smears. But any specific modifications of erythrocytes morphology which are distinctive for one or other disease were not detected.

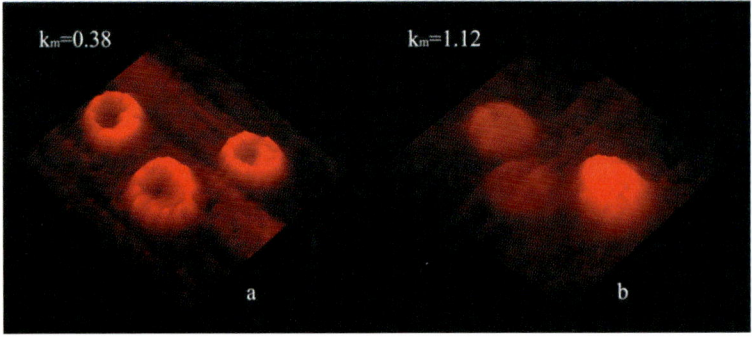

Fig.21. Fragments of native blood smears and their mean sphericity coefficients: health patient (a) and diseased patient (b). The images are obtained using the DHIM.

The development of a disease is always connected with functional disturbance of some sells of an organism. From the other hand, functional disturbance of some cells may cause the development of a disease. The question what is the reason of the observed morphological variations of blood erythrocytes is open and needs further investigations.

All observed morphological modifications of erythrocytes are nonspecific and show the increase in their sphericity coefficients. The increase of the sphericity coefficient corresponds to the decrease of the surface area of the erythrocyte under its given volume. The morphological modifications related to the decrease in the erythrocyte surface area cause a decrease in the functionality with respect to the oxygen supply of tissues and organs.

3D model of erythrocyte was considered to quantitatively estimate the effect of morphological modifications of blood erythrocytes on their functionality with respect to the oxygen transfer [37]. As it is known that the sphere has the minimum surface area under the given volume. So, functionality of spherocytes is minimal. On the assumption that the erythrocyte functionality with respect to the oxygen transport is linearly related to the surface area, we can introduce the morphological functionality factor of the erythrocyte f which is the ratio of the erythrocyte surface area to the surface area of the spherocyte with the same volume:

$$f = \frac{S}{S_0}, \tag{68}$$

where $S_0 = 4\pi r_0^2$ (the surface area of the sphere).The functionality factor $f = 1$ for spherocyte and for other shapes $f > 1$.

Our calculations show that increase in the sphericity coefficient k (transformation of the biconcave erythrocyte to the spherocyte) leads to a significant decrease in the surface area and in functionality of erythrocytes. Theoretically sphericity coefficient k takes on values in the range from 0.1 to 1.155. The first value corresponds to the very biconcave disk erythrocyte, and the second one to a spherocyte erythrocyte (sphere). The morphological functionality factor of an erythrocyte f linearly decreases when the sphericity coefficient k increases.

For example, at $k = 0,4$ (biconcave disk erythrocyte), the morphological functionality factor of the erythrocyte $f = 1.7$; at $k = 1.0$ (flat disk erythrocyte) it is equal $f = 1.2$, and for a spherocyte $f = 1$. It means that the surface area of the flat disk erythrocyte is smaller than the surface area of the biconcave disk erythrocyte by the factor 1.4. The surface area of the spherocyte is smaller than the surface area of biconcave disk erythrocyte by a factor 1.7.

The total surface area of all the blood erythrocytes S_g is equal to the product of the number of erythrocytes in blood N and the mean surface area of erythrocytes S_m:

$$S_g = NS_m. \tag{69}$$

This value determines the oxygen capacity of blood. Taking into consideration equations (68), we can express S_g as:

$$S_g = S_o Nf_m, \tag{70}$$

where S_0 is a constant, which does not depend on erythrocytes shapes; f_m is the mean morphological functionality factor of blood which is simple mean of morphological functionality factors of erythrocytes. Evidently, the oxygen capacity depends on both the number of erythrocytes in blood and shapes of erythrocytes. A variation in the erythrocyte morphology in the direction of the increase of the sphericity coefficient (the transformation of medically normal biconcave disks to flat disks and spherocytes) leads to a decrease in the total surface area of all the erythrocytes in blood by a factor of about 1.4 and 1.7 correspondently. So, morphological modifications of blood erythrocytes decrease the oxygen capacity of blood. That is equivalent to the corresponding decrease of the oxygen capacity of the blood because of decrease in the number of normal (biconcave) erythrocytes in blood by the factors 1.4 and 1.7, correspondently, that can occur in the case of a hemorrhage. The modifications of the erythrocyte shapes detected in the blood smears of the investigated patients (even if the number and volume of erythrocytes and hemoglobin content correspond to the norm) cause the decrease in the oxygen capacity of blood which is equivalent to the loss of one third of all blood erythrocytes (moderate hemorrhage).

A decrease in the oxygen capacity of blood causes hypoxia (the state that emerges upon the insufficient oxygen supply of organs and tissues of an organism). In the medical literature, the blood hypoxia is only related to a decrease in the erythrocyte mass (the number of erythrocytes), and the morphological modifications of erythrocytes as the reason of hypoxia is not considered. Though, it is obvious that morphological modifications of blood erythrocytes result in essential decrease of the functionality of separate erythrocytes and the oxygen capacity of whole blood.

The hypoxia caused by various reasons (including modifications of the erythrocyte morphology) induces reactions aimed at maintaining homeostatic. If the adaptive response of an organism is insufficient, functional damages and structural changes are initiated.

Our investigations of blood erythrocytes morphology are pioneer. They put a lot of questions. The mechanism and reasons of the observed erythrocytes modifications have to be studied. But the necessity of 3D erythrocyte morphology examination under general medical investigation of the patients is out of doubt.

4.2 Ozone therapy influence on 3D morphology of blood erythrocytes

Lately ozone therapy is used in medical practice for normalization of supplying organs and tissues of an organism with oxygen. In our case ozone therapy was used for patient with neurosensor hardness of hearing [26, 29, 31]. The vitality of the problem is determined by its social importance, low efficiency of the existing methods of treatment. Hardness of hearing belongs to the diseases which are caused by negative influence of civilization. As to the data of the International Health Organization more than 450 millions people (that is about 7% of the population of our planet) are suffering with different hearing pathologies.

According to the modern understanding of the problem, the main reason of the neurosensor hardness of hearing is deficiency of oxygen on any level of hearing sensors with posterior negative influence on the functions of the hearing system.

60 patients (men and women) with neurosensor hardness of hearing of different genesis were investigated. The ages of patients were from 18 to 67. All patients took the course of mono ozone therapy. The course of treatment consisted of 5 intravenous injections of 200 *ml* of ozonized physiological solution. Concentration of ozone in the solution was 0.48 mg/liter. Ozonizing of the physiological solution was performed using the device of the Russian firm "Medozone". Untreated blood smears of the patients were prepared before and just after injections of ozonized physiological solution.

All patients passed the audio testing after the course of ozone therapy. In all cases the positive medical effect was achieved.

Figure 22 shows images of blood smears of a patient before and after the injection of the ozonized physiological solution obtained using the classical optical microscope. It is seen that some variations in erythrocytes morphology happened; but 2D images does not allow seeing the character of these variations.

Figure 23 shows 3D images of the blood smears of the patient before and after the injection of the ozonized physiological solution obtained using the DHIM.

It is seen that erythrocytes of the patient before the treatment have flat disk shapes (Fig.23a). After the injections of the ozonized physiological solution the shapes of the erythrocytes are changed. The erythrocytes have the shapes of biconcave disks (Fig.23b). It is apparent that flat disk shape of blood erythrocytes which does not correspond to the norm is the main reason of oxygen deficiency, which causes neurosensor hardness of hearing. Injections of ozonized physiological solution result in normalization of erythrocyte morphology, in increasing their functional possibilities, and, consequently, in the positive medical effect. Such results were obtained for all patients. Shapes of biconcave disk erythrocytes of healthy donors were not changed after the injections of ozonized   physiological solution.

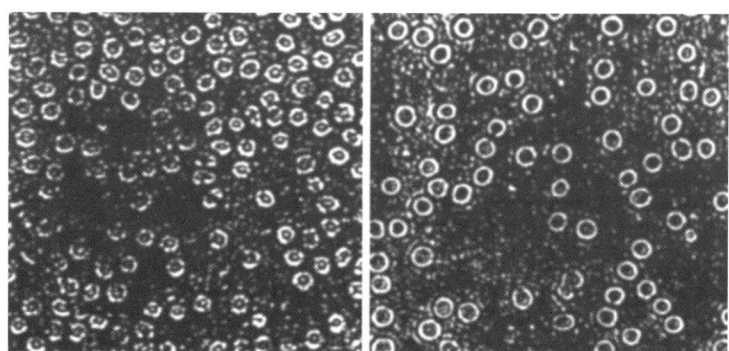

Fig.22. Images of blood smears of a patient before (at the left) and after (at the right) the injection of the ozonized physiological solution. The images are obtained using a usual optical microscope.

Of course, the lack of oxygen in tissues caused by the erythrocytes modifications affect negatively a whole organism and can be the reason of different pathologies. The hardness of hearing is only the manifest pathology. The reason of the observed morphological modifications of

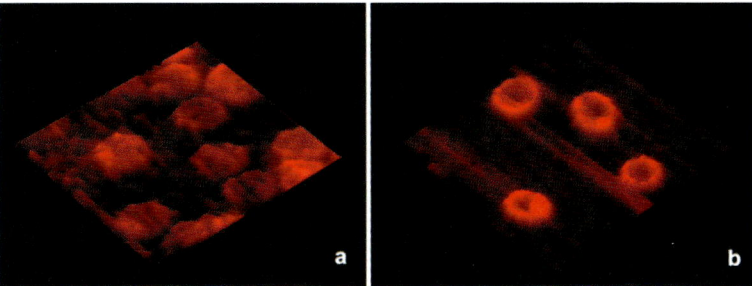

Fig.23. Fragments of blood smear of the patient before (a) and after (b) the injection obtained using the DHIM.

blood erythrocytes of the patients is unknown. It is obvious that ozone therapy influences erythrocyte morphology, but the reason of the morphological modifications and the mechanism of the influence have to be studied.

4.3 Blood erythrocytes in hematological diseases

It is known that morphological modifications of blood erythrocytes are observed in hematological diseases. A lot of degenerative shapes of erythrocytes were detected in blood smears of patients in hemolytic diseases. They resist classification (Fig.24). Such morphological modifications of blood erythrocytes decrease their functionality and are the reason of hypoxic damages of tissues and organs in the diseases. Interesting results were obtained in the case of investigation of a patient in hereditary blood disease such as sickle-cell disease [25]. It is known that sickle-cell disease is an inherited blood disorder that affects red blood cells. People with sickle cell disease have erythrocytes that contain mostly hemoglobin S, an abnormal type of hemoglobin. Hemoglobin S distorts the shape of red blood cells, especially when there is low oxygen. The distorted erythrocytes are shaped like sickles. The functionality of such erythrocytes is reduced. These sickle-shaped cells deliver less oxygen to the body's tissues and organs. When sickle-shaped cells block small blood vessels, less blood can reach that part of the body. Tissue that does not receive a normal blood flow eventually becomes damaged. This is what causes the complications of sickle-cell disease.

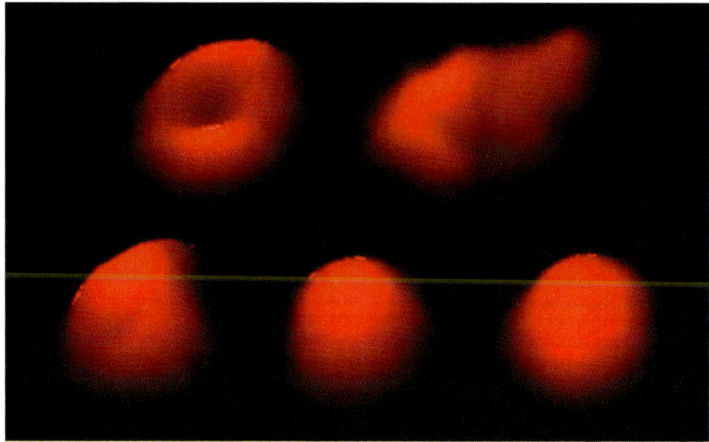

Fig. 24. Modifications of erythrocytes morphology of the patient in hemolytic disease. 3D images are obtained using the DHIM.

There is currently no universal cure for sickle-cell disease. The sickle-cell disease, usually presenting in childhood, occurs more commonly in people from tropical regions where malaria is or was common. People with the sickle-cell disease are more resistant to malaria. The first images of the sickle-shaped erythrocytes were obtained using the electron microscopy method which needs preliminary treatment of a specimen.

Application of the digital holographic interference microscope made it possible to obtain 3D images of native untreated erythrocytes in sickle-cell disease and to observe their transformation.

Figure 25 shows images of blood smears of a patient in the sickle-cell disease (a woman, 65 years old,) before and after decreasing oxygen content in blood (the local hypoxia was created by clamping the patient's finger).The images were obtained using the optical microscope.

One can see that in the state of the local hypoxia essential morphological modifications of erythrocytes shapes happened. Though, the images do not allow making conclusion on 3D shapes of the erythrocytes. Application of the DHIM makes it possible to obtain the obvious information on transformation of erythrocyte shapes in the condition of low oxygen content in blood. Figure 26 shows the dynamics of erythrocyte transformation obtained using the DHIM. It was found

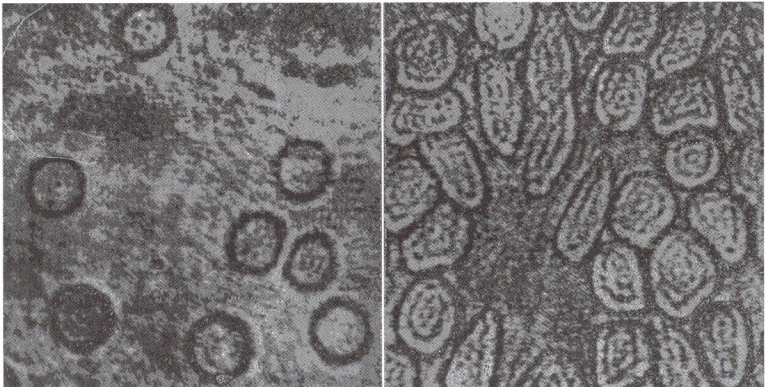

Fig. 25. Fragments of blood smears of the patient in sickle-cell disease before (at the left) and after (at the right) the decrease oxygen content in blood. Images are obtained using the optical microscope.

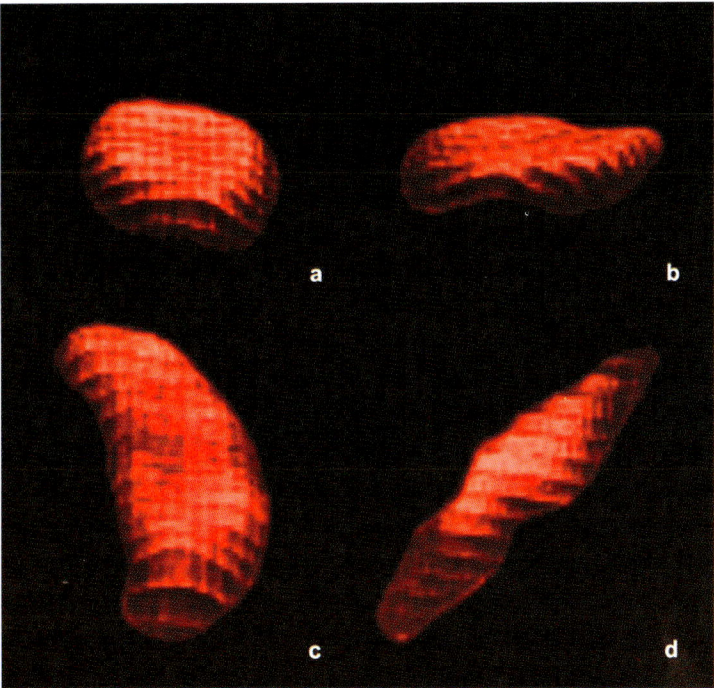

Fig. 26. Changes of erythrocyte shapes of the patient in sickle-cell disease under decreasing oxygen content in blood (a, b, c, d). 3D images are obtained using the DHIM.

that the transformation of erythrocyte shapes occurred for 5 minutes. One can see the intermediates of erythrocyte shape modifications.

4.4 Blood erythrocytes of pregnant women and newborns

The most frequent reason of neonatal mortality and different pathologies of newborns is prenatal hypoxia. Hypoxia is not a disease but the result of different pathologies of mother's and fetus's organisms. Enough supply of a fetus with oxygen is provided by uterine and placental blood circulation intensity, placental barrier penetration and functionality properties of mother's and fetus's erythrocytes. As it was shown above, different diseases can influence blood erythrocytes morphology, and the morphological modifications can be the reason of different hypoxia pathologies of a mother and a child.

Diabetes mellitus is a widespread chronic disease. Progressive increase in the sickness rate put diabetes mellitus on the level of the leading problem of medicine which needs thorough investigation. Deficiency of insulin is responsible for the disease. This results in the disorder of the carbohydrate and protein metabolism. Protein synthesis decreases; glucose content in blood increases. Glucose damages tissues and cells of an organism, and, thus, erythrocytes. Morphological modifications of blood erythrocytes of a mother in diabetes mellitus can be the reason of the fetal hypoxia. This induced us to investigation of 3D morphology of blood erythrocytes of pregnant women (recently confined) without hematological pathology and their newborn infants [35, 38].

Blood smears were prepared in a maternal hospital specializing in diabetes mellitus. 54 persons (41 women and 13 newborns) were investigated. The group of investigated women consisted of pregnant women suffering from carbohydrates metabolic derangement (without hematological pathology; aged from 19 to 36). The control group consisted of 10 pregnant women in norm and 10 healthy non-pregnant women. New born infants born by the women with carbohydrates metabolic derangement and by the healthy women were investigated in an hour and in the third day after deliveries. Untreated blood smears of peripheral blood on glass substrates, prepared according to the standard procedure, were used in our experiments.

It was detected that only 5 from 41 women had biconcave erythrocytes which correspond to the medical norm. Flat disk erythrocytes dominated in blood smears of other 36 women. Specific changes of 3D erythrocytes morphology of patients in diabetes metabolic derangement have not been detected. But the results have turned to be rather unexpected. It has been found that erythrocytes of majority of pregnant women with and without the pathology, and, of woman from the control group did not correspond to the medical norm. Such morphological modifications of erythrocytes can be the reason of different hypoxia pathologies which decrease functional possibility of a human organism.

Correlation between erythrocytes shapes of a mother and her child has not been detected in our experiments. Erythrocytes of all investigated neonatal patients had flat disk shapes.

Figure 27 shows a fragment of blood smear of a neonatal patient.

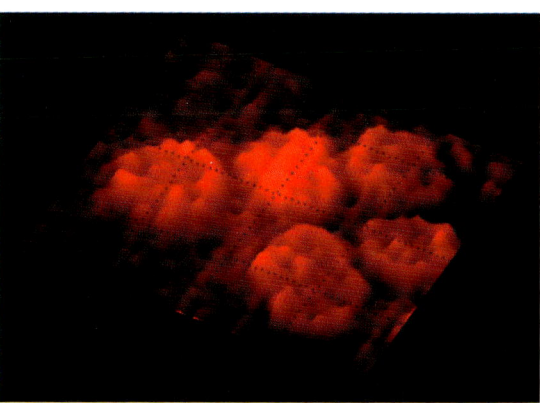

Fig. 27. The fragment of blood smear of a neonatal patient obtained using the DHIM.

Thus, it is possible to conclude that flat disk shape is the norm and the morphological feature of newborns' erythrocytes. As it was considered above, functional possibilities of flat disk erythrocytes are not maximal. The necessary level of the oxygen capacity of blood can be maintained by large number of erythrocytes (see Eq. (69)) and increased content of hemoglobin. Really, as it is known from medical literature, that blood of newborns in first days of their life is characterized by increased number of erythrocytes, increased content of hemoglobin, and presence of young form erythrocytes. This is the reaction of a fetus organism on oxygen deficiency and anoxia in the period of the intrauterine life and delivery. The information concerning erythrocytes shapes of newborns was absent. Now we can say that the morphological types of normal blood erythrocytes of newborns and adults are different. Newborns in first days of their life have flat disk erythrocytes, and crythrocytes of healthy adults are considered to be biconcave disks.

It is know that number of erythrocytes and hemoglobin content begin decreasing in a week after the birth, and these blood parameters are minimal at the age of 2-6 months. This is explained by hemolysis of fetal erythrocytes. Then these parameters begin increasing, and they reach adult values in the age of 12-18. The morphological features of erythrocytes in this period have not been investigated. We think that such investigations will be of great interest.

4.5 *Gamma*-radiation influence on blood erythrocytes

Now in conditions of aggravating environment much attention are paid to investigation of different damaging factors influence on a living organism. In accordance with the modern scientific data, the main targets of different damaging factors influence are plasmatic cell membranes, in particular, membranes of red blood cells. Changes of the elastic properties of the cells membranes must result in morphology changes of erythrocytes. Ionizing radiation influence is of particular interest because the modern life and technical progress are impossible without the radiation. The radiation is used in technique and medicine.

The influence of *gamma*-radiation on morphology of blood erythrocytes of rats *in vivo* and in vitro in lethal dose was studied in our works [15, 28]. The animal and blood of the animal were exposed to irradiation from the *gamma* source (60 C); the dose of irradiation was lethal $(200\,Gr)$.The results of the experiments have confirmed the sensitivity of the blood erythrocytes to the radiation.

Figure 28 shows the results of the experiments. As one can see, dominate in the smear of arterial blood of the rat before irradiation

(Fig. 28a).The mean sphericity coefficient of the smear is equal to 0.2. The blood erythrocytes preserved their biconcave shape in an hour after irradiation (Fig. 28c), but the mean sphericity coefficient of the erythrocytes has increase to 0.5. So, the radiation caused morphological transformation of erythrocytes in the direction of the sphericity coefficient increase. In the smear of the blood irradiated in *vitro* the total hemolysis of erythrocytes was observed.

In addition, the difference of erythrocyte shapes in arterial blood and venous blood from a tail of the rat was detected. The erythrocytes of arterial blood are biconcave (Fig.28a), whereas the flat disc erythrocytes dominate in venous blood (Fig.28b). It can be supposed that erythrocytes can change their shape in the process of gas exchange.

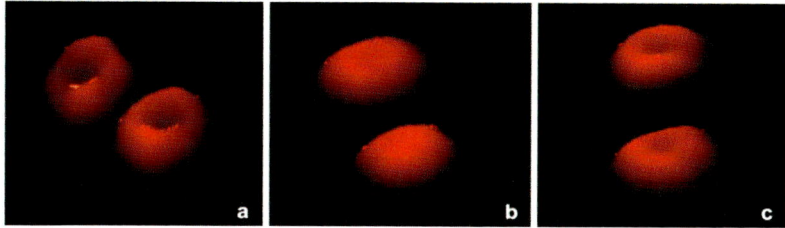

Fig.28. Erythrocytes of arterial (a) and venous (b) blood of a rat. Erythrocytes of arterial blood of the rat after irradiation by *gamma*-radiation in the dose 200Gr *in vivo* (c).

Cancer is a very serious problem. In many countries, more than a quarter of deaths are attributable to cancer. Ionizing radiation (radiation therapy) is used in medical practice as part of cancer treatment. Radiation therapy is the called ionizing radiation to kill cancer cells. Radiation therapy injures or destroys cells in the area being treated by damaging their genetic material, making it impossible for these cells to continue to grow and divide. Radiation damages both cancer cells and normal cells, including blood erythrocytes. There was no information concerning influence of the radiation therapy on morphology of human blood erythrocytes in medical literature. Our experiments with rats have shown that erythrocytes are sensitive to the radiation influence, and morphological modifications of rats' erythrocytes were observed after the irradiation. This allowed us to suppose that morphological modifications of blood erythrocytes can be observed in blood of patient after the course of the radiation therapy. Figure 29 shows the results of study of the radiation therapy influence on blood erythrocytes of a patient in breast cancer (the third stage, woman, 53 years old). The typical erythrocytes of blood smear before and after the course of the radiation therapy are presented. The course of radiation therapy involved 25 session of irradiation; the total dose of irradiation was equal $70\,Gr$.

Biconcave disk erythrocytes which correspond to the medical norm dominate in blood smear of the patient before the course of radiation therapy (Fig. 29a). The mean sphericity coefficient of the blood smear is equal to 0.35. The considerable changes of erythrocytes morphology are observed after the course of the radiation therapy. The biconcave disk erythrocytes became transformed to flat disk erythrocytes; the mean sphericity coefficient of the blood smear is equal to 0.9 (Fig.29b). So, the radiation causes the essential morphological modifications of blood erythrocytes. As it was shown above, such quick transformation of blood erythrocytes results in essential decrease of the oxygen capacity of blood that leads to hypoxia of organs and tissues of the organism. This is an essential damaging factor for a human organism in the course of radiation therapy.

The study of ionizing radiation influence on morphology of blood erythrocytes of rats and patients has shown that erythrocyte morphology is sensitive to ionizing radiation influence. The essential morphological modifications of blood erythrocytes occur. These modifications are nonspecific and go in the direction of sphericity coefficient increase. Such modifications decrease the oxygen capacity of blood and can be the reason of different hypoxic pathologies.

4.6 Conclusions

The digital holographic interference microscope is an effective instrument for 3D visualization of native blood cells and measurement of their morphological parameters.

The proposed method makes it possible to detect pathological modification of erythrocyte morphology and quantitatively characterize the level of their functionality.

The use of the sphericity coefficient for quantitative description of erythrocyte morphology secures the sufficient level of unambiguity and obviousness of the results.

The DHIM experimental study of the 3D morphology of human blood erythrocytes have proved that, in addition to hematological diseases, the

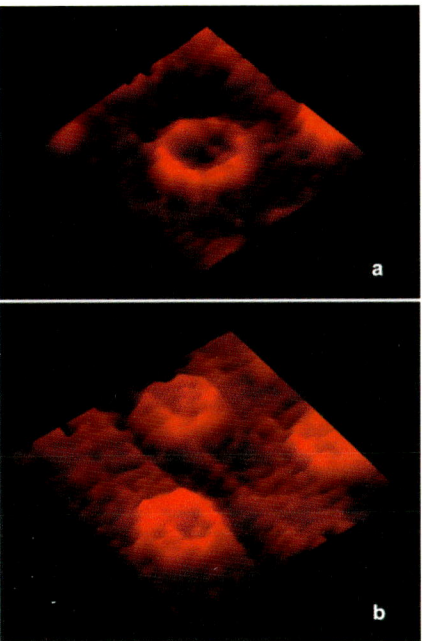

Fig.29. Fragments of blood smears of a patient in breast cancer before (a) and after (b) the course of radiation therapy. The dose of irradiation equals 70 Gr.

morphological modifications of blood erythrocytes are observed in diseases of various genesis and occur under influence of external physical and chemical factors. The morphologic modifications are nonspecific. These nonspecific modifications correspond to increase in the erythrocyte sphericity coefficients and lead to a significant decrease in the oxygen capacity of blood which can be the reason of hypoxia of tissues, organs and, hence, functional and structural changes of organism.

So, it is proved that erythrocyte 3D morphology reflects and determines the state of a living organism and the level of its biological response on different external factors influence.

Chapter 5

DHIM Study of Thin Transparent Films

Lately the scientific and technical industry connected with thin films production is extensively developing. In the process of thin film deposition it is necessary to evaluate the quality of thin film surface and to measure sicknesses of the films. For such investigation optical microscopes are widely used. But they allow only 2D imaging, quantitative measurements are impossible. Application of microinterferometers has some advantages due to the possibility of films thickness measurements. High accuracy of thickness measurement has been achieved for optically nontransparent coatings. But the known optical microinterferometers do not make it possible to obtain good results for transparent thin films (0.1-0.5 μm) on transparent substrates. Till now the electron microscopy method was the single method for thin film morphology investigation. But this method does not make it possible to carry out quantitative measurements. So, the problem of thin transparent films investigation has not been solved.

Thin films are phase microscopic objects. That makes it possible to use the DHIM for thin films investigation. In this chapter we present our pioneer results of application the DHIM for investigation of thin films morphology.

Plastics are widely used now. Lenses made of such materials, along with well known advantages, have shortcomings such as hyper plasticity, and a low value of hardness. Moreover, plastics lose its optical properties under influence of environment, especially solar radiation. Thin transparent films are used as protective coating for optical elements made of plastic. Most of the plastic optical elements are acrylic. Acryl loses its optical properties under the influence of solar radiation. So, protective coatings must protect the surface of such optical elements from the damaging influence of UV-radiation and thermal influences. High transparence in the optical range, high firmness and chemical inertness makes *AlN* the perspective chemical compound for use on the plastic goods which are used for operation in the conditions of environment influence, especially under ultraviolet radiation influence. The technology of deposition of *AlN* -coating on acryl substrates by vacuum-arc method is working out in the National Scientific Center "Kharkov Institute of Physics and Technology". Thin AlN-coatings on acrylic substrates were used as the objects under study in our experiments [39, 40].

Figure 30 shows a fragment of an interferogram of the edge of AlN-coating on the acrylic substrate (a) and the reconstructed 3D image of the fragment. The measured thickness of the coating is about $0.58 \, \mu m$.

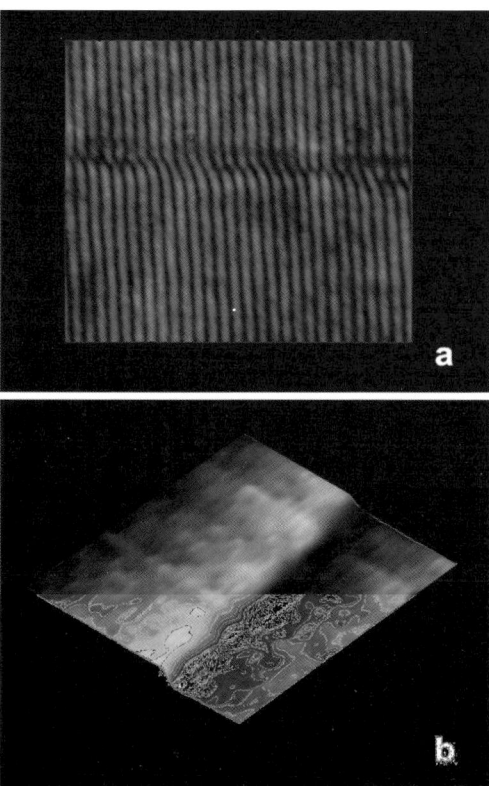

Fig.30. The fragment of the interferogram of the AlN-coating edge (a) and the 3D image of the fragment (b). The length of the edge is about $10 \, \mu m$.

Figure 31 shows the fragments of the surface of *AlN* -coating deposited on acryl substrate with a deposition defect.

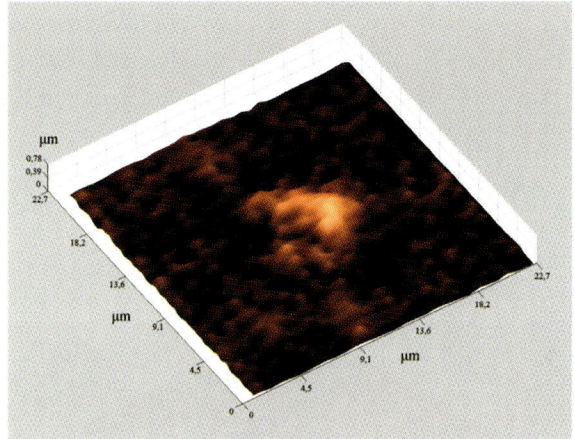

Fig.31. The fragment of the surface of AlN-coating with the deposition defect.

Figure 32 shows the fragment of the surface of *AlN* -coating deposited on acryl substrate with the artificially created scratch.

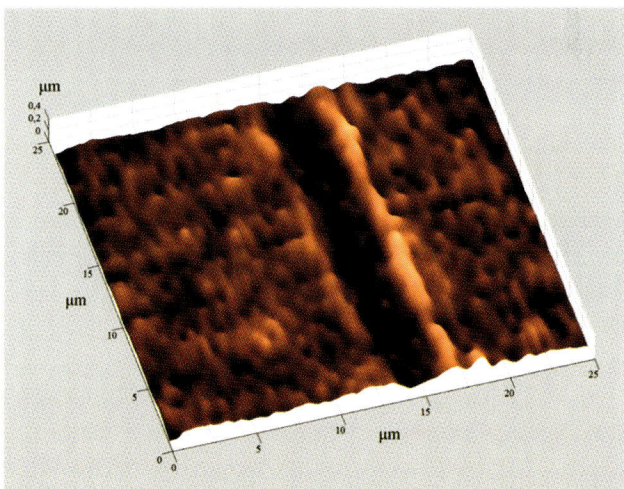

Fig.32. The fragment of the surface of the AlN-coating with the artificially created scratch.

DHIM application makes it possible to determine the character of the defects (convex upwards or concave downward) and to measure the sizes of the defects. So the maximum height of the surface defect (Fig.31) is about $0.7\,\mu m$. The width of the scratch is about $2\,\mu m$ and the depth is about $0.1\,\mu m$ (Fig.32). The obtained images show that film surface (except the places of the defects localization) is plain and homogeneous.

Figure 33 shows the fragment of the surface of AlN-coating on acryl substrate after UV-radiation influence.

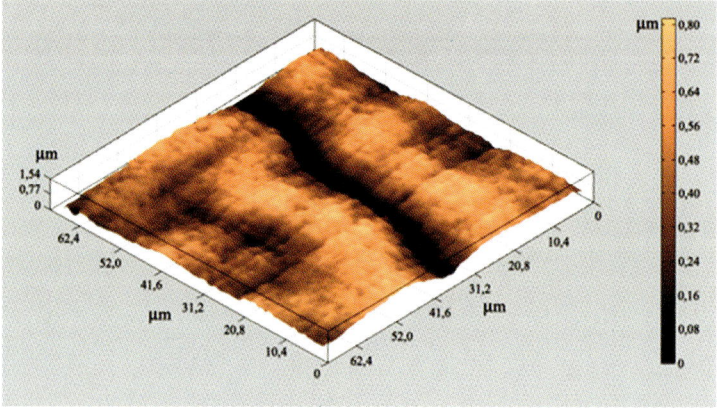

Fig.33. The fragment of the surface of AlN-coating on the acryl substrate damaged by UV-radiation.

Acryl without the protective coating loses its transparence .under the influence of UV-radiation. The *AlN* -coating protects acryl against the damage. Though, when the critical dose of irradiation is achieved, cracks, detachments, and wrinkles on the surface of the coatings appear. One can see the crack on the surface of the coating after the influence of *UV* -radiation (Fig.33). The depth of the crack is about $0.8\,\mu m$. This means that the crack penetrate deep into the coating and into the substrate. So, the coating failed to protect the acryl substrate against damaging influence of *UV* -radiation. It is also seen that the surface of the coating is damaged; it is wrinkled, and uneven.

Figure 34 shows the surface of the *AlN* -coating on the acryl substrate after thermal damage.

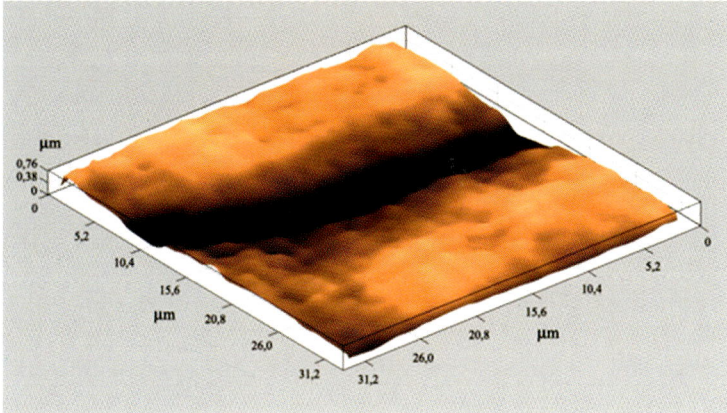

Fig. 34. Fragment of the surface of AlN- coating deposited on acryl substrate after thermal damage.

So, the DHIM can be successfully used for transparent thin film morphology investigation. It combines the possibility of 3D imaging with the possibility of the measurement of the geometrical parameters of the films (thickness, sizes of defects, roughness).

Conclusions

Combining the holographic methods with the methods for digital image processing has solved the problem of 3D visualization of phase microscopic objects. Creation of the digital holographic interference microscope allowing 3D real time imaging of phase microscopic objects and measurement of their morphological parameters has led to a new stage in light microscopy. For the first time 3D imaging of untreated microscopic objects has become possible. We presented pioneer results of DHIM application for investigation of medical and technical phase microscopic objects. These results obviously testify that 3D imaging allows obtaining new and very important information on the microscopic objects under study. So, the digital holographic interference microscopes can find application in medicine, biology and technique.

References

1. Abbe, E. (1873).Beitrage zur Theorie des Mikroskops und der mikroskopischen Wahrnehmung. *Arch.mikrosk.Anat.Entwichlungsmech.*, 9, pp. 413-468.
2. Born, M., Wolf, E.(1973) *Principles of Optics* (Nauka, Moscow, Ru).
3. Ellis, G.W. (1966). Holomicrography: transformation of image during reconstruction a posteriory. *Science.*, 154 , pp. 1195-1196.
4. Franson, M. (1961) *Phase-Contrast and Interference Microscope* (Gosizdat Fiz.-Mat.Lit., Moscow, Ru).
5. Gabor, D. (1949).Microscopy by reconstructed wavefront. *Proc.Roy.Soc.*, A 197 ,p. 454.
6. Ginsburg, V.M., Stepanov, B.M. (1981) *Holographic Measurements* (Radio I Svyas, Moscow, Ru).
7. Leith ,E.N., Upatnieks, J.(1962).Reconstruction wavefronts and communication theory. *J. Opt. Soc. Am.*, 52, pp. 1123.
8. Leith, E.N., Upatnieks, J. (1965).Microscopy by wavefront reconstruction. *J. Opt. Soc. Am.*, 55 ,pp. 569.
9. Lighten, F., Ostenberg, G. (1966).Holographic microscopy. *Nature*, 211, pp. 282-283.
10. Marquet, P., Rappaz, B., Magistretti, P. J., Cuche, E., Emery, Y., Colomb, T., Depeursinge, C.(2005).Digital holographic microscopy: a noninvasive contrast imaging technique allowing quantitative visualization of living cells with subwavelength axial accuracy. *Opt. Lett.*, 30**,** pp. :468–470.
11. Novitsky, V.V., Ryazantseva, N.V., Stepovaya, E.A., Bystritsky, L.D., Tkachenko, S.B.(2003) *Atlas. Clinical Erythrocyte Pathomorphosis* (State University Publishers, Tomsk, Ru).
12. Novitsky, V.V., Ryazantzeva, N.V., Stepovaya, E.A., Shevtzova, N.M., Miller, A.A., Zaitzev, B.N., Tishko, T.V., Titar, V.P., Tishko, D.N.(2008) *Theory and practice of erythrocyte microscopy*(Pechatnaya . Manufactura, Tomsk, Ru).
13. Ostrovsky, Yu. I., Butusov, M.M., Ostrovskaya, G.V. (1977) *Holographic Interferometry*, (Nauka, Moscov,Ru).
14. Popescu, G., Ikeda, T., Best, C. A. (2005).Erythrocyte structure and dynamics quantified by Hilbert phase microscopy. *J. Biomed. Opt.*, 10, pp. 503-508.
15. Paranich, A.V., Tishko, T.V., Titar, V.P.(2002). Application of holographic interference microscopy method for study of gamma-radiation influence on rat blood erythrocytes under irradiation *in vivo* and *in vitro* . *Vestn. Khark .Nats. Univ., Ser. Radiofiz. Electron.*, 544,pp. 40-45 (in Russian).

16. Pluta, M.(1988) *Advanced Light Microscopy*(Elsevier, New York).

17. Salsbury, A.J., Clarke, J.(1967).New method for detecting changes in the surface appearance of human red blood cells. *J. Clin. Pathol.*, 20,p.603.

18. Safronov, G.S., Tishko, T.V., Garagataya, A.M.(1983).Polarization filtering in holographic microscopy. *Ukrainsk. Fizich. Zh., 10*, pp. 1472-1475 (in Russian).

19. Safronov, G.S., Tishko, T.V., Garagataya, A.M. (1984).Holographic Polarization Microscope. *Prib. Tech. Eksp.*, 2, p. 237(in Russian).

20. Safronov, G.S., Tishko, T.V. (1985).Obtaining contrast images of phase microscopic objects by wavefronts summing. *Ukrainsk. Fizich. Zh.*, 30, pp. 334-337(in Russian).

21. Safronov, G.S., Tishko, T.V. (1985).Holographic interferometry of phase microscopic objects. *Ukrainsk. Fizich. Zh.*, 30,pp.994-997(in Russian).

22. Safronov, G.S., Tishko, T.V. (1987). Phase-contrast Holographic Microscope. *Prib. Tech. Eksp.*, 2, p. 249.

23. Safronov, G.S., Tishko, T.V. (1989).Forming difference holograms using orthogonal linear polarized waves. *Ukrainsk. Fizich. Zh.*, 34, pp.72-73 (in Russian).

24. Tishko, T. V., Titar, V. P., Panfilov, D. A., Tishko, D. N. (1998).Application of the holographic interferometry method for determination of shapes of human blood erythrocytes. *Vestn. Khark .Nats. Univ.,Ser Biol. Vestnik.*, 2 (1)**,**pp.107-111(in Russian).

25. Tishko, T.V., Titar, V.P., Nekrasov, V.I., Shpachenko, O.V.(2000).Determination of erythrocytes pathology in sickle-cell diseases fy the holographic interferometry method, *Pros. 14th International Scientific and Practical Conference on the Use of Lasers inMedicine and Biology*, p.16 (in Russian).

26. Tishko, T.V., Titar, V.P., Barchotkina, T.M. (2000). Application of the holographic interference microscopy method for study ozone-therapy influence on shape of human blood erythrocytes. *Vestn. Khark .Nats. Univ.Ser.Boifiz. Vestnik.*, 497, pp.103-107 (in Russian).

27. Tishko, T. V., Titar, V.P. (2001).Holographic microscopy. Methods, devices, applications, *Proc. 3rd International Workshop on Laser and Fiber-Optical Networks Modeling*, pp. 162–167.

28. Tishko, T.V., Titar, V.P.(2002).Investigation of gamma-radiation influence on rat blood erythrocytes *in vivo* and *in vitro* by holographic interfernce method, *Pros. 4th International Workshop on Laser and Fiber-Optical Network Modeling*, LFNM, pp.266-268.

29. Tishko, T.V., Titar, V.P., Barchotkina, T.M., Tishko, D.N.(2003).Application of the holographic interference microscope for investigation of ozone therapy influence on blood erythrocytes of patients *in vivo*, *Pros.1st International Conference on Advanced Optoelectronics and Optics,* CAOL, pp. 6-8.

30. Tishko,T.V., Titar, V.P., Tishko, D.N. (2004). Application of the holographic phase contrast method for 3-D imaging of phase microobjects, *Proc. 6-th International Conference on Laser and Fiber-Optical Networks Modeling,*LFNM, pp. 254-256.

31. Tishko, T.V., Titar, V.P., Barchotkina, T.M., Tishko D.N.(2004).Application of the holographic interference microscope for investigation of ozone therapy influence on blood erythrocytes of patients *in vivo, Proc SPIE,* 5582,pp.119-123.

32. Tishko,T. V., Titar, V. P., Tishko, D. N.(2005).Holographic methods of three-dimensional visualization of microscopic phase objects.*J. Opt. Technol.,* 72(2), pp. 203-209.

33. Tishko, T.V., Titar, V.P., Tishko, D.N.(2005).Application of the digital holographic microscope for investigation of human red blood cells, *Proc. 2nd International Conference on Advanced Optoelectronics and Lasers,*CAOL, pp.284-287.

34. Tishko, T.V., Titar, V.P., Tishko, D.N. (2006). Application of the digital holographic interference microscope for study of human red blood cells. *Vestn. Khark .Nats. Univ.,Ser. Radiofiz. Electron.*, 10 (712),pp. 52-56 (in Russian).

35. Tishko, T.V., Titar, V.P., Tishko, D.N.(2006).Study of blood eryhthrocytes morphology of pregnant women and newborn infants by the of the digital holographic interfernce microscope. *Vestn. Khark .Nats. Univ., Ser.Biolog.,* 729, pp.281-285 (in Russian).

36. Tishko, T.V., Titar, V.P., Tishko, D.N. (2008). 3D morphology of blood erythrocytes by the method of digital holographic interfernce microscopy,*Proc. SPIE*, 7006, pp**.** 70060O-70060O-9.

37. Tishko, T.V., Titar, V.P., Tishko, D.N., Nosov, K.V.(2008).Digital holographic interfernce microscopy in the study of 3D morphology and functionality of human blood erythrocytes. *Laser Physics.,* 18(4), pp.1-5.

38. Tishko, T.V., Titar, V.P., Tishko, D.N. (2008).Using the digital holographic interfcrence microscope for investigation of blood erythrocytes morphology of pregnant women and newborn infant, *Pros. 4th International Conference on Advanced Optoelectronics and Lasers,*CAOL, pp. 171-173.

39. Tishko, T.V., Tishko, D.N., Titar, V.P.(2008).Digital holographic interference microscopy of phase microscopic objects, *Proc. 4th European Microscopy Congress,*1,pp.285-286.

40. Tishko, D.N., Tishko, T.V., Titar, V.P (2009).Using digital holographic microscopy to study transparent thin films. *J. Opt. Technol.,*76, pp.147-149.

41. Zernike, F.(1934). Diffraction theory of the knife-edge test and its improved form, the phase-contrast method. *Royal Astronomy Society Monthly Notices,* 94,pp. 377-384.

Bibliography

Abbe, E. (1873).Beitrage zur Theorie des Mikroskops und der mikroskopischen Wahrnehmung. *Arch.mikrosk.Anat.Entwichlungsmech.*, 9, pp**.** 413-468

Born, M., Wolf, E.(1973) *Principles of Optics* (Nauka, Moscow, Ru)

Ellis, G.W. (1966). Holomicrography: transformation of image during reconstruction a posteriory. *Science.,* 154 , pp. 1195-1196.

Franson, M. (1961) *Phase-Contrast and Interference Microscope* (Gosizdat Fiz.-Mat.Lit., Moscow, Ru).

Gabor, D. (1949).Microscopy by reconstructed wavefront. *Proc.Roy.Soc.*, A 197 ,p. 454

Ginsburg, V.M., Stepanov, B.M. (1981) *Holographic Measurements* (Radio I Svyas, Moscow, Ru).

Leith ,E.N., Upatnieks, J.(1962).Reconstruction wavefronts and communication theory. *J. Opt. Soc. Am.*, 52, pp. 1123.

Leith, E.N., Upatnieks, J. (1965).Microscopy by wavefront reconstruction. *J. Opt. Soc. Am.*, 55 **,**pp. 569.

Lighten, F., Ostenberg, G. (1966).Holographic microscopy. *Nature*, 211, pp. 282-283.

Marquet, P., Rappaz, B., Magistretti, P. J., Cuche, E., Emery, Y., Colomb, T., Depeursinge, C.(2005).Digital holographic microscopy: a noninvasive contrast imaging technique allowing quantitative visualization of living cells with subwavelength axial accuracy. *Opt. Lett.,* 30**,** pp.468–470.

Novitsky, V.V., Ryazantseva, N.V., Stepovaya, E.A., Bystritsky, L.D., Tkachenko, S.B.(2003) *Atlas. Clinical Erythrocyte Pathomorphosis (* State University Publishers, Tomsk, Ru).

Novitsky, V.V., Ryazantzeva, N.V., Stepovaya, E.A., Shevtzova, N.M., Miller, A.A., Zaitzev, B.N., Tishko, T.V., Titar, V.P., Tishko, D.N.(2008) *Theory and practice of erythrocyte microscopy(* Pechatnaya . Manufactura, Tomsk, Ru).

Ostrovsky, Yu. I., Butusov, M.M., Ostrovskaya, G.V. (1977) *Holographic Interferometry*, (Nauka, Moscov,Ru).

Popescu, G., Ikeda, T., Best, C. A. (2005).Erythrocyte structure and dynamics quantified by Hilbert phase microscopy. *J. Biomed. Opt.*, 10, pp. 503-508.

Paranich, A.V., Tishko, T.V., Titar, V.P.(2002). Application of holographic interference microscopy method for study of gamma-radiation influence on rat blood erythrocytes

87

under irradiation *in vivo* and *in vitro* . *Vestn. Khark .Nats. Univ., Ser. Radiofiz. Electron.,* 544,pp. 40-45 (in Russian).

Pluta, M.(1988) *Advanced Light Microscopy*(Elsevier, New York).

Salsbury, A.J., Clarke, J.(1967).New method for detecting changes in the surface appearance of human red blood cells. *J. Clin. Pathol.*, 20,p.603.

Safronov, G.S., Tishko, T.V., Garagataya, A.M.(1983).Polarization filtering in holographic microscopy. *Ukrainsk. Fizich. Zh., 10*, pp. 1472-1475 (in Russian).

Safronov, G.S., Tishko, T.V., Garagataya, A.M. (1984).Holographic Polarization Microscope. *Prib. Tech. Eksp.*, 2, p. 237(in Russian).

Safronov, G.S., Tishko, T.V. (1985).Obtaining contrast images of phase microscopic objects by wavefronts summing. *Ukrainsk. Fizich. Zh.*, 30, pp. 334-337(in Russian).

Safronov, G.S., Tishko, T.V. (1985).Holographic interferometry of phase microscopic objects. *Ukrainsk. Fizich. Zh.*, 30,pp.994-997(in Russian).

Safronov, G.S., Tishko, T.V. (1987). Phase-contrast Holographic Microscope. *Prib. Tech. Eksp.*, 2, p. 249.

Safronov, G.S., Tishko, T.V. (1989).Forming difference holograms using orthogonal linear polarized waves. *Ukrainsk. Fizich. Zh.*, 34, pp.72-73 (in Russian).

Tishko, T. V., Titar, V. P., Panfilov, D. A., Tishko, D. N. (1998).Application of the holographic interferometry method for determination of shapes of human blood erythrocytes. *Vestn. Khark .Nats. Univ.,Ser Biol. Vestnik.*, 2 (1)**,**pp.107-111(in Russian).

Tishko, T.V., Titar, V.P., Nekrasov, V.I., Shpachenko, O.V.(2000).Determination of erythrocytes pathology in sickle-cell diseases fy the holographic interferometry method, *Pros. 14th International Scientific and Practical Conference on the Use of Lasers inMedicine and Biology*, p.16 (in Russian).

Tishko, T.V., Titar, V.P., Barchotkina, T.M. (2000). Application of the holographic interference microscopy method for study ozone-therapy influence on shape of human blood erythrocytes. *Vestn. Khark .Nats. Univ.Ser.Boifiz. Vestnik.*, 497, pp.103-107 (in Russian).

Tishko, T. V., Titar, V.P. (2001).Holographic microscopy. Methods, devices, applications, *Proc. 3rd International Workshop on Laser and Fiber-Optical Networks Modeli*ng, LFNM, pp. 162–167.

Tishko, T.V., Titar, V.P.(2002).Investigation of gamma-radiation influence on rat blood erythrocytes *in vivo* and *in vitro* by holographic interfernce method, *Pros. 4th International Workshop on Laser and Fiber-Optical Network Modeling,* LFNM, pp.266-268.

Tishko, T.V., Titar, V.P., Barchotkina, T.M., Tishko, D.N.(2003).Application of the holographic interference microscope for investigation of ozone therapy influence on blood erythrocytes of patients *in vivo, Pros.1st International Conference on Advanced Optoelectronics and Optics,* CAOL, pp. 6-8.

Tishko,T.V., Titar, V.P., Tishko, D.N. (2004). Application of the holographic phase contrast method for 3-D imaging of phase microobjects, *Proc. 6-th International Conference on Laser and Fiber-Optical Networks Modeling,*LFNM, pp. 254-256.

Tishko, T.V., Titar, V.P., Barchotkina, T.M., Tishko D.N.(2004).Application of the holographic interference microscope for investigation of ozone therapy influence on blood erythrocytes of patients *in vivo, Proc SPIE,* 5582,pp.119-123.

Tishko,T. V., Titar, V. P., Tishko, D. N.(2005).Holographic methods of three-dimensional visualization of microscopic phase objects.*J. Opt. Technol.,* 72(2), pp. 203-209.

Tishko, T.V., Titar, V.P., Tishko, D.N.(2005).Application of the digital holographic microscope for investigation of human red blood cells, *Proc. 2nd International Conference on Advanced Optoelectronics and Lasers,*CAOL, pp.284-287.

Tishko, T.V., Titar, V.P., Tishko, D.N. (2006). Application of the digital holographic interference microscope for study of human red blood cells. *Vestn. Khark .Nats. Univ.,Ser. Radiofiz. Electron.,* 10 (712),pp. 52-56 (in Russian).

Tishko, T.V., Titar, V.P., Tishko, D.N.(2006).Study of blood eryhthrocytes morphology of pregnant women and newborn infants by the of the digital holographic interfernce microscope. *Vestn. Khark .Nats. Univ., Ser.Biolog.,* 729, pp.281-285 (in Russian)'

Tishko, T.V., Titar, V.P., Tishko, D.N. (2008). 3D morphology of blood erythrocytes by the method of digital holographic interfernce microscopy,*Proc. SPIE,* 7006, pp**.** 70060O-70060O-9.

Tishko, T.V., Titar, V.P., Tishko, D.N., Nosov, K.V.(2008).Digital holographic interfernce microscopy in the study of 3D morphology and functionality of human blood erythrocytes. *Laser Physics.,* 18(4), pp.1-5.

Tishko, T.V., Titar, V.P., Tishko, D.N. (2008).Using the digital holographic interference microscope for investigation of blood erythrocytes morphology of pregnant women and newborn infant, *Pros. 4th International Conference on Advanced Optoelectronics and Lasers,*CAOL, pp. 171-173.

Tishko, T.V., Tishko, D.N., Titar, V.P.(2008).Digital holographic interference microscopy of phase microscopic objects, *Proc. 4th European Microscopy Congress,*1,pp.285-286.

Tishko, D.N., Tishko, T.V., Titar, V.P (2009).Using digital holographic microscopy to study transparent thin films. *J. Opt. Technol.,*76, pp.147-149.

Zernike, F.(1934). Diffraction theory of the knife-edge test and its improved form, the phase-contrast method. *Royal Astronomy Society Monthly Notices,* 94,pp. 377-38Amsden, A. A. and Harlow, F. H. (1970). The SMAC method: A numerical technique for calculating incompressible fluid flows, Los Alamos Sci. Lab. Rep. No. LA-4370.

Index